Modernizing China's Military

Opportunities and Constraints

Keith Crane • Roger Cliff • Evan Medeiros
James Mulvenon • William Overholt

Prepared for the United States Air Force
Approved for public release; distribution unlimited

PROJECT AIR FORCE

The research reported here was sponsored by the United States Air Force under Contract F49642-01-C-0003. Further information may be obtained from the Strategic Planning Division, Hq. USAF.

Library of Congress Cataloging-in-Publication Data

Modernizing China's military : opportunities and constraints / Keith Crane ... [et al.].
  p. cm.
  "MG-260."
  Includes bibliographical references.
  ISBN 0-8330-3698-X (pbk. : alk. paper)
  1. China—Armed Forces—Reorganization. 2. China—Military policy. I.
Crane, Keith, 1953–

UA835.M63 2005
355.3'0951—dc22

2004030176

The RAND Corporation is a nonprofit research organization providing objective analysis and effective solutions that address the challenges facing the public and private sectors around the world. RAND's publications do not necessarily reflect the opinions of its research clients and sponsors.

RAND® is a registered trademark.

Cover design by Stephen Bloodsworth

Published 2005 by the RAND Corporation
1776 Main Street, P.O. Box 2138, Santa Monica, CA 90407-2138
1200 South Hayes Street, Arlington, VA 22202-5050
201 North Craig Street, Suite 202, Pittsburgh, PA 15213-1516
RAND URL: http://www.rand.org/
To order RAND documents or to obtain additional information, contact
Distribution Services: Telephone: (310) 451-7002;
Fax: (310) 451-6915; Email: order@rand.org

Preface

This report is designed to help the U.S. Air Force assess the resources the government of the People's Republic of China is likely to spend on its military over the next two decades. The study projects future growth in government expenditures as a whole and the military in particular, evaluates the current and likely future capabilities of China's defense industries, and compares likely future expenditure levels with recent expenditures by the United States and the U.S. Air Force.

The research reported here was sponsored by AF/XOX and PACAF/CC and conducted within the Strategy and Doctrine Program of RAND Project AIR FORCE. It is part of the RAND Corporation's ongoing research on China and China's military establishment. The study builds on previous RAND Project AIR FORCE work:

- Roger Cliff, *The Military Potential of China's Commercial Technology*, MR-1292-AF, 2001
- Mark Burles and Abram Shulsky, *Patterns in China's Use of Force: Evidence from History and Doctrinal Writings*, MR-1160-AF, 2000
- Erica Strecker Downs, *China's Quest for Energy Security*, MR-1244-AF, 2000
- Richard Sokolski, Angel Rabasa, and C. R. Neu, *The Role of Southeast Asia in U.S. Strategy Toward China*, MR-1170-AF, 2000
- Mark Burles, *Chinese Policy Toward Russia and the Central Asian Republics*, MR-1045-AF, 1999
- Daniel L. Byman and Roger Cliff, *China's Arms Sales: Motivations and Implications*, MR-1119-AF, 1999

- Zalmay Khalilzad, Abram N. Shulsky, Daniel Byman, Roger Cliff, David T. Orletsky, David A. Shlapak, and Ashley J. Tellis, *The United States and a Rising China: Strategic and Military Implications,* MR-1082-AF, 1999
- James C. Mulvenon and R. H. Yang (eds.), *The People's Liberation Army in the Information Age,* CF-145-CAPP/AF, 1999.

RAND Project AIR FORCE

RAND Project AIR FORCE (PAF), a division of the RAND Corporation, is the U.S. Air Force's federally funded research and development center for studies and analyses. PAF provides the Air Force with independent analyses of policy alternatives affecting the development, employment, combat readiness, and support of current and future aerospace forces. Research is performed in four programs: Aerospace Force Development; Manpower, Personnel, and Training; Resource Management; and Strategy and Doctrine.

Additional information about PAF is available on our web site at http://www.rand.org/paf.

Contents

Figures

Tables

Summary

The Problem

A number of U.S. analysts and policymakers have raised concerns about the potential for China to mount a serious strategic challenge to the United States in Asia, especially in the western Pacific, sometime in the course of the next two decades. These concerns are based on China's expanding economy: The rapid economic growth of the past three decades has dramatically increased the resources the Chinese government has available to devote to military spending. Recent double-digit percentage increases in officially reported defense budgets indicate the degree to which China's growing economic base has permitted the Chinese government to increase the resources it expends on the military.

For these concerns to become reality by 2025, a number of intervening events must take place. First, the economy will have to continue to grow. Second, the government will have to be able to extract revenues from the economy for military expenditures. Third, to increase military spending dramatically, the government will have to ward off competing pressures for higher expenditures on pensions, health care, and education, and more public investment in infrastructure. And fourth, China's defense industries will have to be able to produce the sophisticated weaponry that China would need to seriously challenge U.S. forces.

Approach

The purpose of this study is to assess future resource constraints on and potential domestic economic and industrial contributions to the ability of the Chinese military to become a significant threat to U.S. forces by 2025. We conducted this assessment by answering the following questions:

1. What will be the likely shape and size of the Chinese economy over the next two decades?
2. What types of constraints will the Chinese government face in terms of drawing on increased economic output for spending on the military?
3. What problems will the military face and what possibilities will it have in terms of purchasing the goods and services it desires from the Chinese defense industry?
4. How will these constraints and opportunities shape the capabilities of the Chinese armed forces over the next two decades?

We developed answers to these questions using a variety of information sources and analytical techniques. Chinese statistical data, analyses of the Chinese economy, and a model of the Chinese economy were used to address the question of economic growth and size. The rich literature discussing tax, social, and fiscal policies in China was married with a statistical analysis of Chinese spending and the economic analysis to address the questions of budgetary constraints. To evaluate the health of China's defense industries, we engaged in an extensive analysis of open-source Chinese and English language information on these industries and interviews with knowledgeable industry specialists to conduct a thorough analysis of institutional reforms in China's defense industries, including contracting procedures. The results of this analysis are presented in Chapter Five, China's Defense Industry. Estimates of the current and future size of Chinese military expenditures drew on newly available Chinese language primary sources concerning defense budgeting and local expenditures, Chinese statistical material, including provincial statistical yearbooks, and a military spending forecasting model developed for this study.

Economic Growth Is Destined to Slow, but Output Still Will Triple by 2025

Measured at purchasing power parity (PPP) exchange rates, an exchange rate estimated by comparing the purchasing power of local currencies for the same basket of goods across countries, China's economy is already the second largest in the world, lagging behind only that of the United States. At market exchange rates, its economy is the sixth largest, lagging behind the United States, Japan, Germany, the United Kingdom, and France.

The Chinese economy is destined to become even larger. We project that China's economy will grow at an average annual rate of 5 percent through 2025, more than tripling in size (see pp. 44–45). The projected rates of growth, 7 percent per year through 2010, gradually declining to 3 percent per year in 2025, are substantially lower than the average annual rate of growth of 8.7 percent reported for the past quarter century. However, the average rate of 5 percent per year over the forecast period is more plausible than forecasts employing rates of growth from the recent past. First, the Organization for Economic Cooperation and Development (OECD), World Bank, and Chinese economic research institutes all argue persuasively that official growth rates have been exaggerated. Second, and more important, China faces a host of constraints that will serve to slow growth over the course of the next two decades: stagnation and eventual decline in the labor force, a fall in domestic savings as the population ages, a slowdown in growth in exports and industrial output because of market saturation, weaknesses in the financial sector, and problems in agriculture and rural areas.

To compare economic output and hence potential resources available for military expenditures with that of other countries, China's GDP must be converted from renminbi, the domestic currency, into a common denominator such as dollars. Economists use two types of exchange rates to make these conversions: the market exchange rate and the PPP exchange rate. Neither of these exchange rates is altogether satisfactory for converting China's GDP into dollars over time. On the one hand, the Chinese government keeps the market exchange rate undervalued by fixing or pegging the renminbi against the dollar, thereby imparting a downward bias to GDP converted at the market exchange rate. If, as forecast, China's economy continues to grow rapidly, its market exchange rate will appreciate in real effective terms over the next two decades, increasing the value of China's GDP in dollar terms above and beyond increases in output measured in constant prices in renminbi. On the other hand, the PPP exchange rate is heavily influenced by the low cost of goods and services produced with local labor, such as housing, personal services, and basic foodstuffs. As incomes and wages rise, so will relative prices of these goods and services. Furthermore, as Chinese households become wealthier, they are likely to shift their consumption toward consumer durables and other tradable goods and services, the prices of which are determined by international markets, not local factors intrinsic to China. These changes will lead to the real effective *depreciation* of the PPP exchange rate. This depreciation will serve to reduce China's rate of growth in dollar GDP below that of renminbi GDP.

Because of the deficiencies of both exchange rates for comparing economic output and because of these two contrary trends in exchange rates, we chose to measure future Chinese output in dollars using a combination of both exchange rates. We explicitly projected likely future changes in both the market and purchasing power parity exchange rates. We used the projected market exchange rates to convert tradable goods and services into 2001 dollars and the projected PPP exchange rates to convert nontraded goods and services into 2001 dollars. Employing this technique, we project that China's GDP will be $9.45 trillion in 2001 dollars by 2025. At that time, China's economy will be slightly smaller than the U.S. economy was in 2003. Assuming the U.S. economy grows at an average annual rate of 3 percent over the next two decades, China's economy would be about half the size of the U.S. economy in 2025 (see pp. 46–48).

China's Government Faces Increasing Pressures to Increase Social Spending

Since economic reform began in the late 1970s, China has been undergoing a series of dramatic changes in its systems of taxation and provision of public services. Public expenditures and revenues as a share of GDP dropped sharply between 1979 and the mid-1990s as the Chinese government stopped financing investments by state-owned enterprises and the government struggled to replace tax revenues from state-owned enterprises with revenues from taxes on the market-oriented non-state sector. Simultaneously, the central government attempted to control tax revenues and expenditures at the provincial level. During this period of change, the central government's share of taxes and expenditures fell sharply: Tax revenues as a share of GDP reached a low of 10.7 percent in 1995, an extraordinarily small share of output. Pressures to increase government expenditures on education, research and development, and the military led to a reversal of this trend in the mid-1990s as the central government instituted tax reforms that raised collections. As of 2002, central government revenues had risen to 18.5 percent of GDP, a substantial increase.

Political pressures for increased public services are so strong and the current provision of education, health care, and pensions so inequitable and dysfunctional that the Chinese government will be compelled to continue to increase the share of output spent on these government services and transfers in the coming decades. In addition, the Chinese government has a number of latent liabilities, such as unfunded commitments for pensions and the continued need to recapitalize state-owned banks as they write off the very large bad debts of state-owned enterprises. When these liabilities are recognized, they will boost government debt substantially, imposing substantial future debt servicing costs on the government. Financing these increased expenditures will be difficult. Taxes and government expenditures as a share of GDP are already above the average in medium-income developing countries. Further increases in tax rates threaten to incite political unrest—tax revolts are already common in rural areas.

Despite the difficulty in raising government revenues as a share of GDP, we project that political pressures will drive future Chinese government spending on pensions, health care, education, and interest payments on rising government debt from 7.5 percent of GDP in 2000 to 15.5 percent in 2025, an increase of 8.0 percentage points (see p. 89).

Deconstructing China's Defense Budgets

The official Chinese defense budget includes only a portion of the total defense budget. Although it includes most personnel, operations and maintenance, and equipment costs, the following items are excluded:

- Foreign weapons procurement
- Expenses for paramilitaries (People's Armed Police)
- Nuclear weapons and strategic rocket programs
- State subsidies for the defense-industrial complex
- Some defense-related research and development
- Extra-budget revenue that goes to the military (*yusuanwai*).

We have reestimated China's defense budget by including estimates of these omitted items. To the official budget of 185.3 billion renminbi in 2003 ($22.4 billion), we have added estimates of Chinese imports of military equipment ($3.6 billion), provincial support to national defense ($1.18 billion), and paramilitary expenses ($3 billion). Finally, we have assumed that defense-industrial subsidies and R&D funding are bounded by the totals listed in the national budget, and could not exceed $3.1 billion and $4.3 billion, respectively. Using this combination of data and assumptions, we estimate that China's total defense expenditures ran between $31 to $38 billion in 2003, or 1.4 to 1.7 times the official number (see p. 133).

Reforms Hold Promise of Improving Weak Performance of China's Defense Industries

Although China's resources are substantial, the People's Liberation Army (PLA) has frequently not been able to purchase the weapon systems it desires from domestic manufacturers. Over the past 20 years, China's defense industrial base has had great difficulty in producing technologically sophisticated, high-quality weapons systems. This state of affairs appears to be changing. Beginning in spring 1998 during the 9th Meeting of the National People's Congress, the Chinese government initiated a "grand strategy" for improving the technological capabilities of China's defense industries. This strategy has three main elements. The first is selective modernization. China's leaders hope to exploit China's strengths in electronics technology and missiles by concentrating on command, control, communications, computers, intelligence, surveillance and reconnaissance (C4ISR), and accurate strike weapons. The second element of the strategy is civil-military integration. The government has attempted to provide incentives and make organizational changes to capture for military enterprises improvements in efficiency and technological sophistication that state-owned enterprises in computing, shipbuilding, and electronics have made in production for civilian clients. As part of this process, the Chinese government is reforming defense procurement, introducing bidding for some contracts. It has also restructured the defense industries at the enterprise level, breaking the defense industry "companies" into semiautonomous enterprises that are to compete with one another and foreign suppliers. The third element of the grand strategy is to exploit advanced foreign technology. The government has embarked on a program of substantial imports of weapons, equipment, and military technologies from Russia, Israel, and other suppliers.

These strategies appear to have had some success in improving the sophistication and quality of military equipment produced by the information technology, shipbuilding, and aerospace industries. On the civilian side, these industries are manufacturing globally competitive products. A number of new weapons systems, including Chinese destroyers, missiles, and C4ISR systems, have shown marked improvements over past production because they incorporate more sophisticated technologies acquired from production for civilian clients. In contrast, China's military aviation industry continues to underperform China's other defense industrial sectors. Its diversification into commercial production of aviation and non-aviation products has not significantly contributed to the modernization of the industry. The People's Liberation Army Air Force (PLAAF) has had to rely on imports to acquire aircraft that are even somewhat competitive with those flown by the United States.

xxii   Modernizing China's Military: Opportunities and Constraints

Many of the weaknesses of China's defense industrial sector could be overcome in the short to medium term, assuming it does not deviate from the present course of reform and continues to invest in defense production. If the government continues to push for open contracting and takes a tough line on cost overruns, the rate of innovation and quality of weapons systems should continue to improve. However, improvements in efficiency will not happen overnight. It will take time to change management behavior and stimulate innovation, even after new management incentive systems are implemented (see p. 190).

PLA Leadership Perceives United States as Greatest Threat

Threat perceptions and the desire to project power are key drivers of the acquisition of military capabilities. PLA military strategists perceive the United States as posing both an immediate and long-term challenge to Chinese national security interests. Thus, by far the most immediately relevant driver of the PLA's current planning and procurement is the goal of the Chinese leadership and the PLA to reassert control over Taiwan and their concerns about possible U.S. intervention if conflict with Taiwan should ensue. Since the end of the 1990s, PLA reform, modernization, procurement, and training has been heavily—almost totally—focused on preparing for a conflict over Taiwan. Beyond the narrow Taiwan contingency, Chinese military planners and political leaders are decidedly uncomfortable with the U.S. military presence in the world; they fear that the United States can and will use military force whenever and wherever it wants, including in scenarios involving Chinese security interests. A related security concern for Chinese military planners is Japan. Although Chinese political leaders continue to value Sino-Japanese economic relations for their contribution to domestic economic growth, Chinese military strategists remain concerned about the possible rebirth of Japanese militarism and about Japan's military alliance with the United States. Finally, protecting Chinese territorial waters and airspace has long been a primary mission for the Chinese military. Over the past two decades, this mission has been expanded to protecting Chinese claims to parts of the South China Sea.

The PLA seeks to modernize its force structure to provide it with the capabilities to meet these threats and challenges. It has placed particular emphasis on acquiring four categories of capabilities (see pp. 202–203):

1. The capability to respond to both internal and external threats by quickly taking the initiative, preventing escalation, attaining superiority, and resolving the conflict on China's terms
2. The eventual development of a limited power projection capability which would facilitate a sustained sea presence and an area denial capability, although area control is not a high priority for the PLA
3. The ability to conduct short-range preemptive strikes using conventional missiles and air force assets
4. The development of a credible strategic nuclear capability to deter other nuclear powers from using nuclear threats to coerce China or to limit its strategic options, especially during a crisis.

Chinese Military Spending Is Likely to Rise to $185 Billion by 2025

In light of prospective rates of economic growth, pressures for increasing government spending on categories other than defense, and Chinese threat perceptions, what are the possible and likely levels of future military spending? To answer this question, we generated two sets of projections of potential Chinese military expenditures through 2025 in 2001 dollars (Table S.1). Our high-end forecast was based on our assumption that the maximum share of output the Chinese government would be able to spend on defense in the context of current threat perceptions would be 5.0 percent of GDP. This assumption is based on the considerable demands for expenditures on health care and pensions, in particular, that China's rapidly aging population is placing on the government. It also reflects the upper bound of what middle income developing countries have been willing to spend on defense over the past two decades, and our evaluation of the PLA's assessments of threats to

Table S.1
RAND Projections of Chinese Military Spending Through 2025: Combined
Market and PPP Exchange Rates (billions of 2001 dollars)

	2003	2010	2015	2020	2025
Mid-range projection	68.6	91.2	113.7	143.9	185.2
Personnel	48.9	57.8	65.0	73.1	82.2
Operations and maintenance	8.6	15.3	23.0	34.6	51.9
Procurement and R&D	11.1	18.1	25.6	36.2	51.1
Maximum projection	75.6	145.0	207.4	287.3	403.4
Personnel	48.9	84.7	111.5	141.0	178.9
Operations and maintenance	8.3	22.3	39.6	67.1	113.0
Procurement and R&D	18.5	38.0	56.2	79.3	111.4
Ratio between maximum and mid-range projections	1.10	1.59	1.82	2.00	2.18

China at this point in time. Our mid-range projection is based on the assumption that military spending will not rise above our current lower bound estimate of military spending of 2.3 percent of GDP.

Projections were made by major expenditure category in renminbi and then converted into 2001 dollars using projected market or PPP exchange rates, whichever was more appropriate. For example, personnel costs were converted into 2001 dollars using PPP exchange rates while procurement costs were converted at market exchange rates. We argue that this composite approach provides the most accurate comparison between Chinese and U.S. military expenditures.

Both projections yield very substantial sums (see Table S.1). By 2025, our mid-range projection yields spending of $185 billion, a little over 60 percent of the United States' 2003 defense budget. However, 44 percent of these projected expenditures consist of personnel costs; compared to only a quarter of the total budget in the United States. The projection of military spending under the maximum expenditure scenario results in considerably higher numbers: Military spending rises from an estimated $76 billion in 2003 to $403 billion in 2025, at which time Chinese military spending would be a third greater than the 2003 U.S. defense budget (measured in 2001 dollars). However, this projection is truly a maximum in terms of what China is likely to be able to afford. It is based on an assumption that the Chinese leadership would be willing to raise military expenditures to 5 percent of its GDP during a period when political pressures to increase spending on health, education, and pensions—not to mention infrastructure, the environment, and unemployment assistance—will be very strong (see p. 232).

We also projected the potential resources that China may devote to purchasing military assets in the coming two decades. To provide a better sense of the cumulative impact on force structure of the projected defense expenditures on procurement, we compared the projected cumulative totals spent on procurement and research, development, testing, and evaluation (RDT&E) through 2025 with U.S. expenditures on these items in 2001 dollars over the past 22 years (see Table S.2). Military capabilities, especially stocks of weaponry, are the result of cumulative spending over time, not just current spending. This exercise provides a measure of what China may spend on procurement cumulatively over the course of the next two decades.

Table S.2
Potential Future Chinese Military Expenditures on Procurement Compared with U.S. Expenditures

Category	Expenditures (billions of 2001 dollars)	China as a Percent of U.S.[a]
Cumulative U.S. expenditures on RDT&E and procurement 1981–2003	2,712.4	
Maximum cumulative projections of Chinese expenditures on RDT&E and procurement 2003–2025	1,279.7	47.2
Mid-range cumulative projections of Chinese expenditures on R&D and procurement 2003–2025	597.8	22.0
Cumulative USAF expenditures on R&D and procurement 1981–2003	1,039.4	
Maximum cumulative projections of Chinese expenditures on PLAAF procurement 2003–2025	490.4	47.2
Mid-range cumulative projections of Chinese expenditures on PLAAF procurement 2003–2025	229.1	22.0

[a]Because the share of total procurement by the PLAAF is assumed to be the same as that of the USAF, the ratios between total Chinese procurement and PLAAF procurement and the U.S. budget and USAF procurement are the same.

Procurement was appreciable in the high-end case. The cumulative total would be close to half of what the United States spent on military procurement and RDT&E between 1981 and 2003. Under this scenario, no other country outside the United States would rival China in terms of weapons stocks. In the mid-range projection, procurement spending is still appreciable, but in this case, even after 22 years, China's cumulative expenditures on procurement would be only 22.0 percent of what the United States spent between 1981 and 2003 (see p. 235).

As noted above, the PLA is intent on creating a limited power protection capability and being able to respond quickly to threats to Chinese territory. Air assets would play a key role in creating these capabilities. To provide the U.S. Air Force a more tangible measure of what potential future Chinese expenditures might be on air assets, we also provide notional projections of future Chinese spending on research and development and procurement of air assets and compare them to past USAF expenditures in the same categories. The magnitudes are revealing. In our view, the maximum likely expenditures that China would make on RDT&E and procuring weapons and equipment for the PLAAF between 2003 and 2025 would be on the order of $490 billion. That is a very large sum of money. However, the cost in 2001 dollars of the current USAF inventory of weapons and equipment and the associated RDT&E to develop those systems is more than twice this number. Using our mid-range projection of military expenditures, PLAAF expenditures on procurement and RDT&E would run $229 billion in 2001 dollars, 22.0 percent of cumulative USAF expenditures over the past 22 years on the same categories (see p. 237).

How Can We Tell If China Is Straying from the Projected Course?

Throughout our research for this study, we identified and utilized a variety of indicators to assess current and project likely future trends in Chinese military spending and defense industries. We have culled a subset of these indicators that we believe are most salient for tracking likely future changes. By following these indicators, analysts should be able to spot changes in trends that might herald substantial changes in resources allocated to defense. They should also be able to determine whether future military expenditures and capabilities are likely to diverge substantially, up or down, from those projected here.

Key indicators include the following (see pp. 251–253):

- Rates of growth in GDP
- Shares of GDP taken in revenues and expenditures by the government
- Introduction of a national pension program
- Loss of major contracts by traditional military equipment manufacturers through a competitive bidding process
- Closure of poorly performing plants in defense industries
- Changes in official budgetary expenditures in real terms deflated by the GDP deflator
- Changes in the share of GDP accounted for by official military spending
- Changes in the total government budget for research and development.

Acknowledgments

The authors would like to thank Garret Albert, Matt Roberts, and Eric Valko for their assistance with the statistical research, Maggie Marcum for her insight and expertise, Heather Roy and Karen Stewart for their administrative support, and Miriam Polon and Alissa Hiraga for editing the volume. We would also like to thank Murray Scott Tanner and Nicholas Lardy for two very helpful reviews and David Epstein for his perceptive comments.

Acronyms and Abbreviations

ASCM	antiship cruise missile
C4I	command, control, communications, computers, and intelligence
C4ISR	command, control, communications, computers, intelligence, surveillance, and reconnaissance
CASIC	China Aerospace Science and Industry Corporation
CASC	China Aerospace Science and Technology Corporation
CMC	Central Military Committee
COSTIND	Commission on Science, Technology, and Industry for National Defense
CSY	*China Statistical Yearbook*
DWT	dead weight tons
FDI	foreign direct investment
GAD	General Armaments Department
GDP	gross domestic product
GLD	General Logistics Department
GPCR	Great Proletarian Cultural Revolution
IISS	International Institute for Strategic Studies
IMF	International Monetary Fund
IT	information technology
LTG	Lieutenant General
MBI	Machine Building Industry
MR	Military Region
NDIO	National Defense Industry Office
NDSTC	National Defense Science and Technology Commission
NORINCO	Northern Chinese Industries Corporation
NPC	National People's Congress
PAP	People's Armed Police

PBC	People's Bank of China
PLA	People's Liberation Army
PLAAF	PLA Air Force
PLAN	PLA Navy
PPP	purchasing power parity
PRC	People's Republic of China
R&D	research and development
RDT&E	research, development, testing, and evaluation
RMB	renminbi
SETC	State Economic and Trade Commission
SBI	shipbuilding industry
SOE	state-owned enterprise
STECO	Science and Technology Equipment Commission
TSS	Tax Sharing System
USAF	U.S. Air Force
WTO	World Trade Organization
ZJCSD	*Zhongguo junshi caiwu shiyong daquan* [Practical Encyclopedia of Chinese Military Finance]

CHAPTER ONE

Introduction

Will China Become a Serious Military Threat in the Western Pacific?

A number of U.S. analysts and policymakers have raised concerns about the potential for China to mount a serious strategic challenge to the United States in Asia, especially in the western Pacific, sometime in the course of the next two decades. These concerns are based on China's expanding economy: The rapid economic growth of the past three decades has dramatically increased the resources the Chinese government has available to devote to military spending. Recent double-digit percentage increases in officially reported defense budgets indicate the degree to which China's growing economic base has permitted the Chinese government to increase the resources it expends on the military.

For these concerns to become a reality, a number of intervening events must take place. First, the economy will have to continue to grow. Second, the government will have to be able to extract revenues from the economy for military expenditures either through taxation, by borrowing at home or abroad, or by printing money. Third, the government will have to balance competing pressures for higher expenditures on pensions, health care, education, and more public investment in infrastructure against increased military spending. And fourth, China's defense industries will have to be able to produce the sophisticated weaponry that China would need to seriously challenge U.S. forces.

Some factors suggest that China will have the ability to fund and build a modern military. Continued strong growth in the economy and the budget is likely. Chinese industry may develop the technological wherewithal, at least in some industrial branches, to produce modern weaponry. Signs of increasing technological sophistication in Chinese industry abound. The economy has not only enjoyed very rapid rates of growth over the past few decades, it has been transformed. Large inflows of direct foreign investment, massive imports of modern equipment and machinery, and dramatic increases in human capital stemming from improvements in China's educational system and the return of Chinese students from studying abroad have contributed to the creation of a number of modern industrial sectors, especially in information technology. Some of these industries produce key components for modern military technologies, especially aviation, aerospace, and command, control, communications, computers, intelligence, surveillance, and reconnaissance (C4ISR). If China's economy continues to modernize over the coming decades, military equipment and weapons producers are likely to have access to domestically produced components to construct the military equipment and systems needed to narrow the capabilities gap with the United States.

However, a number of factors may prevent the Chinese government from making the expenditures that would be needed to field armed forces that could present a serious challenge. Some economists have concluded that Chinese growth slowed in the second half of the 1990s, in some years very sharply, because of financial problems in state-owned enterprises and state-controlled banks. More modest growth in GDP would constrain increases in Chinese military spending. Demographic changes are likely to have a major impact on budget expenditures. China's population is aging rapidly, and failure to pay pensions is one of the major sources of protest and unrest in Chinese society today. More public pressure for the provision of state-financed pension schemes could divert budgetary funds from the military to social programs. The government also faces demands for increased spending on health and, to a lesser extent, education.

The Chinese government has not been extraordinarily adept at collecting taxes. Most of China's citizens operate in a cash economy and actively seek to avoid paying taxes, often quite successfully. Small businesses, in particular, are good at tax evasion. Because the government has difficulty in taxing flows in the private sector (sales of goods and services and incomes), it has had to continue to rely on taxing the remaining state-owned sector, especially large state-owned enterprises and banks, or physical assets such as property for a substantial share of revenues. As a consequence, the tax burden often falls most heavily on those the state can compel to pay rather than on those who can most easily afford to pay. The burden on peasants, who pay taxes on physical assets like land and livestock, is proportionally higher than on entrepreneurs in urban areas who do not have substantial physical assets. Endemic corruption and the rapidly growing private sector may make it difficult for the Chinese government to effectively raise the tax revenues needed to increase spending to the levels necessary to fund a much more capable military.

The Chinese government also faces the challenge of ensuring that decisions about resource allocation made by policymakers are implemented. Graft is endemic to the Chinese system. At all government levels, a substantial share of tax revenues is siphoned off by government employees. The prevalence of graft both reduces the resources available to the Chinese government and discourages citizens from paying taxes.

The military also faces constraints in terms of its ability to acquire the goods and services it desires. The Chinese military has to contend with competitive markets for management and leadership talent, restricted sources of supply for advanced weaponry, and institutional weaknesses in integrating weapons systems. Although money is helpful in solving these problems, it is a necessary but not sufficient resource for creating a modern military. The Chinese military faces an enormous challenge in the coming decade to transform itself from a massive conscript army focused on defending Chinese territory to an institution capable of projecting power outside China's borders. If China's leadership wishes to accomplish this objective, its military will have to attract and retain a dedicated, innovative officer corps that functions far differently than the current cadre. The Chinese military will have to compete for these individuals with a private sector that has been growing very rapidly and has made a number of entrepreneurs and managers quite wealthy in the process.

On the supply side, China's ability to acquire the full range of equipment and systems needed by a modern military could well be constrained by the deficiencies of China's defense industry and by external restrictions on imports of more capable equipment from foreign suppliers. Currently, the domestic arms industry either is not able or finds it very difficult to produce modern equipment and integrate it into effective weapons systems. The Chinese military will have to revise its procurement processes to induce the defense industry to develop more sophisticated weaponry and to integrate these weapons into highly capable weapon systems. In addition, despite recent improvements in the professionalism of the People's Liberation Army (PLA) and the People's Liberation Army Air Force (PLAAF), both institutions will have to make substantial changes in their operating procedures to be able to use more technologically sophisticated military equipment in an effective manner.

Purpose of This Study

This book assesses future resource constraints on and potential domestic economic and industrial contributions to the ability of the Chinese military to become a significant threat to U.S. forces in the western Pacific by 2025. It addresses the following key questions:

1. What will be the likely shape and size of the Chinese economy over the next two decades?
2. What types of constraints will the Chinese government face in terms of drawing on increased economic output for spending on the military?
3. What problems will the military face, and what possibilities will it have in terms of purchasing the goods and services it desires from the Chinese defense industry?
4. How does the PLA perceive the military challenges facing China? What types of forces and capabilities does it wish to field to respond to these challenges?
5. Faced with these desires, constraints, and opportunities, what resources will the Chinese armed forces likely have at their disposal over the next two decades?

In addition to supplying answers to these questions, we identify indicators that USAF intelligence analysts may wish to track to determine whether the Chinese leadership is channeling greater or fewer resources than expected to creating a military capability that could mount an effective challenge to U.S. forces in the coming decades.

Outline of the Book

Following this introduction, Chapter Two evaluates recent trends in Chinese economic growth, describes the current structure of the Chinese economy, and forecasts likely future economic trends. As part of this analysis, we examine potential constraints on future economic growth, including incipient problems in the banking system, problems in the rural economy, a slowdown in trade and foreign investment, and the rapid aging of China's population. The chapter concludes with a discussion of likely structural changes in the Chinese economy over the coming two decades with a specific focus on changes that would contribute to the development of future military capabilities.

Chapter Three describes and assesses the composition of current and recent Chinese government expenditures and sources of budgetary revenues at the national and provincial levels. We describe changes over time in terms of expenditure patterns and levels by key categories. We also dissect the roles of various levels of government in overall government spending and look at recent and likely future changes in the roles of these different levels in providing various government services. We evaluate recent and likely future changes in taxes and tax revenues. As noted above, the Chinese government has not been extraordinarily adept at collecting taxes. Therefore we identify areas where the Chinese government may be able to extract additional taxes and areas where the tax burden is already contributing to popular discontent and opposition to the government. We also evaluate constraints on the ability of the Chinese government to raise revenues through public borrowing.

Chapter Four evaluates current levels and trends in resources expended on the Chinese military. Chinese military expenditures, like expenditures in the former Soviet Union, are difficult to track. Aside from the official defense budget, various ministries provide additional funds and subsidies. Provincial and local governments chip in through various programs, including paying conscripts to work on local construction projects. In this chapter, we dissect current military spending, using financial and economic statistics. We also employ recent data on provincial spending on the military that have not been previously available. Drawing on these sources, we provide best estimates of expenditures on the Chinese military, broken down by major expenditure categories and major sources of funds.

In Chapter Five, we evaluate likely changes in the capabilities of China's defense industries to provide more-advanced weaponry by 2025 and in the PLA's ability to effectively contract for these products. We assess four branches of the defense industry:

1. Information technology and defense electronics
2. Aviation
3. Aerospace, with an emphasis on missiles and avionics
4. Shipbuilding.

Our assessments of the domestic industries focus on general capabilities, challenges, and opportunities to create a more modern arms industry. Key issues include organization, systems integration, the quality of the component and subsystem industries, the challenges of generating sufficient revenues to make the industry profitable, and purchases and integration of imported components and technologies. This chapter does not include detailed engineering evaluations of particular technologies of importance to the manufacture of military equipment.

Not only will China's defense industries have to create the capacity to build modern weapons, but the PLA will also have to create the institutions and mechanisms to successfully contract for those weapons. To do so, the PLA will have to revise its procurement processes and its operations. In this chapter, drawing on current Chinese and foreign language sources, we assess the challenges the PLA faces in procuring and integrating weapons and command and control systems, either domestically produced or imported from abroad. We conclude by discussing potential changes in operating procedures that would significantly improve the ability of the PLA to contract for and absorb new systems and assessing the likelihood that such changes will be introduced.

Threat perceptions play a key role in decisions on expenditures on the military. Chapter Six briefly reviews the Chinese military's perceptions of the strategic threats and objectives that face the country. Subsequently, we discuss the types of forces that the Chinese military believes will be needed to fulfill its future missions. Our analysis employs Chinese and foreign writings on these subjects.

In Chapter Seven we project likely future levels of Chinese government spending on the military from the point of view of the budget. We approach this task by first projecting overall budget revenues and expenditures. Based on the experiences of other developing Asian countries, we evaluate potential or likely future changes in taxation, including measures to improve taxpayer compliance with the tax code, and the implications for future tax revenues. We then project likely future demands for major nonmilitary spending, including health care, education, state-supported pensions for China's aging population, and investments in public infrastructure, thereby bounding likely funds available for military expenditures. Subsequently, we project two series of likely future military expenditures based on expected growth in GDP and the expected future share of military spending within the total budget. We complete the chapter by comparing projected expenditures in dollars with past expenditure by the U.S. Department of Defense and the U.S. Air Force.

The book concludes with a set of indicators for USAF intelligence analysts designed to indicate whether the Chinese leadership is making a concerted effort to significantly improve the capabilities of the PLA, i.e., whether the Chinese government is attempting to "break out" by dramatically increasing resources devoted to the military and corresponding improvements in capabilities. In contrast to the 1960s and 1970s, when China was a closed society and information was very limited, U.S. military planners now face the difficult task of absorbing and interpreting a flood of information concerning the Chinese military and Chinese society. Much of this information suggests divergent trends, making it difficult for analysts to accurately assess future developments in the PLA. This is especially true when discussing future rates of military equipment procurement. For example, based on plans reported earlier by the Chinese government, a number of analysts had projected more rapid rates of procurement of new aircraft than the PLAAF eventually fielded. In this final chapter we assemble a set of leading indicators that, based on past performance, have foreshadowed future increases in defense spending, future acquisitions of weapon systems, and improvements in the capabilities of the PLA. We have drawn the bulk of these indicators from the data gathered in the course of conducting the research for the book. They constitute a partial checklist for USAF analysts, a checklist that should indicate likely future trends in Chinese military expenditures and improvements in capabilities.

CHAPTER TWO

The Chinese Economy

Introduction

The key determinant of a country's ability to expend resources on its military forces is the size of its economy. From a U.S. military planning perspective, the major question concerning China's economy is its ability to generate military capabilities both in absolute terms and in comparison to the U.S. economy. This chapter first measures China's economic output. It then assesses the potential for China's economy to continue to grow rapidly over the course of the next three decades. It concludes with projections through 2025 of the future size of China's economy.

The Current Size of the Chinese Economy

China's Economy Compared to That of the United States

Gross domestic product (GDP) is a standard measure of the annual economic output of a country. Aggregate demand (GDP plus net imports of goods and services) determines the resources available to a country for consumption, investment, and government expenditures, including expenditures on the military. GDP is measured in domestic prices and the domestic currency—in the case of China, renminbi (RMB).

To make cross-country comparisons, economic output has to be converted into a common denominator or currency. Two standard methods are employed to compare GDP across countries. One way is to convert the value of GDP into a common currency such as the dollar using the average annual market exchange rate, which is computed by averaging daily exchange rates for the year. A second way is to convert GDP using *purchasing power parity* (PPP) exchange rates. Purchasing power parity exchange rates are determined by comparing the cost of purchasing a common basket of goods and services in one country with its cost in another, again using a common currency. The U.S. dollar is often chosen as a basis for comparison because of its ubiquity and the size and international importance of the U.S. economy.

In the case of many developing countries, including China, the use of market exchange rates to measure GDP or per-capita incomes introduces a downward bias in comparison with developed countries. This bias stems from the lower relative cost of important basic goods and services consumed by households and from the "weakness" of currencies in most developing countries. By definition, wages and incomes are lower in developing countries than in developed countries. Consequently, many important components of the household consumption basket are relatively cheaper. In particular, labor-intensive goods and services like handicrafts, housing, haircuts, dry cleaning, and health care that are consumed locally and that cannot be traded internationally tend to be relatively cheaper in countries with lower wages than in developed countries.

Economic and financial forces tend to make market exchange rates in developing countries "weak," imparting another downward bias to estimates of living standards as measured in dollars or foreign currencies of wealthier countries. Exchange rates help equilibrate a country's external financial flows; financial flows therefore play a major role in determining the rate of exchange. In general, investors, foreign or domestic tend to view developing-country assets as riskier than equivalent assets in more developed countries because economic growth is often more erratic and the risk of default on the part of governments and businesses tends to be higher. Investors therefore demand higher rates of return on equivalent developing country assets to compensate for the greater perceived risk. As a result, all else being equal, local currencies tend to be less attractive than more established currencies such as the dollar.

In the case of China, the official exchange rate policy also imparts a downward bias to comparisons of economic output based on the

official rate because it appears to have kept the renminbi undervalued. China's central bank, the People's Bank of China, has kept the
exchange rate of the renminbi against the dollar constant at 8.28 for
the past several years, even though China has been running current
account surpluses and attracting large inflows of direct foreign investment, factors that under a floating exchange rate would have resulted in
the appreciation of the renminbi. The People's Bank of China pegs the
renminbi against the dollar to provide a stable environment in which
exporters and importers can operate. China is able to maintain this rate
using a combination of central bank intervention and restrictions on
capital and current account convertibility. If the renminbi were stronger, the value of China's GDP and living standards as measured in dollars would be higher.

Because market exchange rates fluctuate, sometimes quite widely,
they introduce problems in comparing countries' economies over time.
And because currencies are financial assets, market exchange rates are
heavily influenced by investor sentiment. As sentiment changes, the
value of the currency changes as well. For example, if domestic or foreign investors believe a government is allowing its budget balance to
deteriorate, the exchange rate tends to depreciate as investors sell government bonds. If the exchange rate depreciates sharply enough, the
size of the economy as measured in foreign currency diminishes. In
some instances, the exchange rate effects may be so strong that the size
of the economy measured in, for example, dollars, may shrink even
though the economy is growing. Conversely, if the exchange rate appreciates, increases in GDP as measured by market exchange rates in
dollars will be exaggerated.

Because of the problems that market exchange rates introduce in
comparing economic output across countries, economists frequently
prefer to use PPP exchange rates for comparisons. because PPP exchanges rates provide a better picture of the purchasing power and
standard of living of the average citizen in a country. However, they,
too, pose problems.[1] The rates are calculated by conducting price sur-

---

[1] For a detailed discussion of the deficiencies of PPP rates in general and current attempts to
calculate and collect country statistics, see Ryten, 1998.

veys of both consumer and producer goods in the reference country and the country of interest. A typical market basket for a consumer in the country of interest, in this case China, is compared with the cost of a similar market basket in the United States. The choice of goods in the survey and the weights given those goods have enormous implications for the calculation of the exchange rate. In the case of China, the "basket" is likely to be heavily weighted toward food, especially rice, low-cost clothing, and housing, which are relatively cheap compared to their costs in the United States. If the market basket of a typical U.S. consumer, which is more heavily weighted to automobiles, vacations, and consumer electronics, were to be substituted for that of an average Chinese consumer, the differences in costs would be much smaller as would the PPP exchange rate. Moreover, a great number of judgment calls enter into the calculation of PPP exchange rates. It is not always possible to find exact matches between countries in terms of goods. Finally, the surveys, generally undertaken under the auspices of the World Bank or other international financial institutions, are conducted relatively infrequently for most countries, often only once in five years. Consequently, once a PPP exchange rate is calculated, it is usually adhered to for a number of years during which time it is not adjusted for changes in relative prices.

In contrast, market exchange rates are available on a real time basis. Standard methodologies exist for generating average market exchange rates on a daily, monthly, or annual basis, eliminating the measurement problems that plague estimates of PPP exchange rates. As an economy becomes more integrated into the global economy, prices of tradable goods are determined to an increasing degree by world market forces and tend to become much more equal across countries. As prices become more similar and the country becomes more integrated into global financial markets, the disadvantages of using market exchange rates for international comparisons decrease.

Dollar Estimates of the Size of the Chinese Economy
By either measure, PPP or market exchange rates, China has an economy large enough to sustain a substantial level of military expenditures in comparison with most countries in the world, excluding the United

States. Although in many respects China remains a very poor country, with an average per-capita income of $960 per annum at market exchange rates in 2002, its gross domestic product (GDP) looms large. In 2002, GDP ran $1,237 billion at market exchange rates, making China's economy just slightly larger than Italy's (Figure 2.1). By this measure, China ranks sixth in the world behind the United States, Japan, Germany, the United Kingdom, and France.

According to the World Bank, China's GDP at purchasing power parity exchange rates was $5,027 billion in 2001 (Figure 2.2). U.S. GDP was $9,781 billion in the same year.[2] Extrapolating to 2002 using official increases in GDP, China's economy was 54 percent of the U.S. economy in 2002. As can be seen, the exchange rate used makes an enormous difference: China's economy is more than four times bigger when measured using PPP exchange rates than it is when market exchange rates are used.

Figure 2.1
China's GDP at Market Exchange Rates Compared to Other Large Economies, 2002

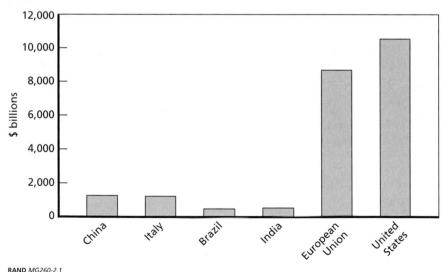

RAND *MG260-2.1*

---

[2] World Bank, 2004.

Figure 2.2
China's GDP Measured at Purchasing Power Parity Exchange Rates Compared to Other Large Economies, 2002

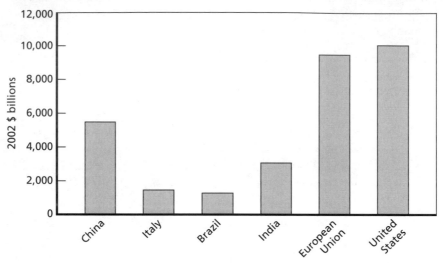

How accurate are these PPP estimates? As noted above, estimates of PPP exchange rates are highly sensitive to the number of items used to make cross-country comparisons, consumption basket weights, and the year in which the comparison is made as well as the quality of the price data and the comparability of goods and services in terms of quality and retail venue. Consequently, estimates can vary widely. In 1995, the World Bank reviewed a number of estimates of China's dollar GDP for 1985 or 1986 and computed its own estimate for 1986.[3] Although the range of estimates was very wide, varying by a factor of 10, only two of those estimates, the Bank's and that of Kravis, were true PPP estimates.[4] The Bank's estimate was a little more than half (54 percent) of the previous estimate by Kravis. The authors of the World Bank report argue that by dint of a larger range of products covered and more detailed price surveys, their estimate was superior to that of Kravis, who used 1975 data.

---

[3] Ren and Chen, 1995, p. 6.

[4] Kravis, 1981.

Their estimates also differ from current World Bank estimates: Their estimate for the PPP exchange rate for 1986 was 27 percent higher than that implied by the current World Bank PPP exchange rate for China. Based on their estimate, China's economy would have been about 26 percent of the U.S. economy in 1986 as opposed to the 21 percent implied by the more recent World Bank figure—or one-quarter higher. In light of significant improvements in Chinese statistics on prices and household expenditures and shifts in relative prices since 1986, we believe that the World Bank's 2001 data are superior. Although the problems of estimating PPP exchange rates are large, we argue that this estimate is the best estimate currently available and that credible alternative estimates of China's economy at PPP exchange rates would not greatly exceed the figure used here.

Measured either by market or by PPP exchange rates, the U.S. economy remains substantially larger than China's, although the difference is much larger using market exchange rates. In 2002, the Chinese economy was only 11.8 percent of the U.S. economy at market exchange rates; at purchasing power parity exchange rates it was 54.3 percent (Figure 2.3). On a per-capita basis, GDP in China was only

Figure 2.3
China's Per Capita and Total GDP as a Share of That of the United States, 2002

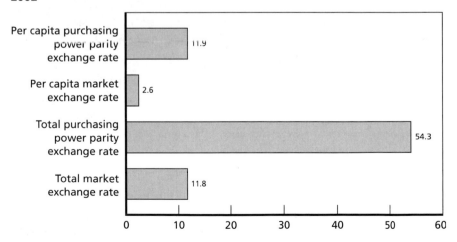

SOURCE: Calculated from data from International Monetary Fund, *World Economic Outlook*, Washington, D.C., April 2004, Appendix.

RAND *MG260-2.3E*

2.6 percent of U.S. levels in 2002 at market exchange rates; at purchasing power parity exchange rates, it ran 11.9 percent of U.S. levels. The extraordinary differences in relative size between the two measures is due primarily to differences in prices of non-tradable goods and services—goods and services that tend to be relatively cheaper in China than in the United States.

Despite the very different pictures presented by these two measures, neither measure is "wrong": Both have their uses when comparing the size of different economies or differences in per-capita incomes. From an external perspective, comparisons of GDP at market exchange rates are often the most useful. They better reflect the purchasing power of households and businesses on international markets. GDP at market exchange rates also provides a better check on borrowing capacity and the role of the country in international financial markets than does GDP at purchasing power parity exchange rates. In contrast, purchasing power parity exchange rates tend to provide a better measure of real differences in living standards.

Measures of GDP and Military Spending

Both exchange rates have their uses for comparing military expenditures across countries. Purchasing power parity exchange rates provide better dollar measures of the resource cost of personnel in developing country militaries. Because wage rates and costs of food, clothing, and housing are often much lower in developing countries than in the United States, the use of market exchange rates tends to underestimate the comparative value of resources devoted to personnel in developing country armed forces. However, a substantial share of more sophisticated military equipment purchased by developing-country militaries, including China's, is imported from abroad or incorporates components such as electronics, diesel engines, and aircraft frames that are manufactured from materials and parts sold at world market prices. Hence, when calculating domestic currency expenditures devoted to procurement, converting expenditures to dollars at market exchange rates may better capture the true cost of these items, especially because military equipment is not used in computations of PPP exchange rates by the World Bank. Because we are interested in questions pertaining to both personnel and procurement, this study marries these two approaches for military expenditures, providing estimates and forecasts of military spending in 2001 dollars using a synthesis that involves both types of exchange rates.

Chinese Economic Growth

Overall Determinants of Chinese Economic Growth

The story of Chinese economic growth since the adoption of a series of market-oriented economic policies that began in 1978–1979 has been well documented.[5] Since the reforms, China's economy has grown very rapidly: Between 1979 and 1997, growth in GDP averaged close to 10 percent per year, according to official statistics. Although these growth rates are impressive, in many ways, China has belatedly followed in the footsteps of Japan, South Korea, and Taiwan, which also enjoyed extended periods of rapid growth in the post–World War II period. Taiwan's economy grew at an average annual rate of 9.3 percent between 1961 and 1989, and South Korea reported average annual rates of growth of 8.1 percent between 1970 and 1991. Japan, Malaysia, Singapore, and Thailand have all enjoyed long periods of very rapid economic growth during the post–World War II period as well.

The same overall factors that drove economic growth in these economies have driven growth in China. These include the following:

- Substantial increases in fixed capital assets resulting from high rates of investment
- Large increases in labor productivity driven by a shift of labor from low-productivity jobs in agriculture to higher-productivity occupations in manufacturing and services
- Large improvements in the stock of human capital through education
- Substantial increases in total factor productivity made possible by the adoption of new technologies

---

[5] See, for example, Overholt, 1993; Wang and Yao, 2001; and World Bank, 1997.

- Better utilization of comparative advantage through the expansion of trade in goods and services and increased factor flows.[6]

The relative importance of these factors differs across countries. Gregory Chow finds that increases in capital were a much more important source of growth in China than in Taiwan, whereas increases in total factor productivity, often viewed as a measure of the contribution of technological change to growth, have been more important in Taiwan.[7] The reliance on increases in capital to spur growth is somewhat troubling because of China's historical problems with efficiently allocating investment—or at least investment funded by the state.

During this period of rapid growth, the importance of a number of specific factors fostering economic growth has waxed and waned. Below, we evaluate some of the contributions of individual economic sectors in spurring China's economic growth and describe how the importance of these sectors has changed over time. We conclude with a discussion of those factors that are likely to constrain or drive future growth in China's economy.

Sources of and Constraints on Future Economic Growth: The Rural Economy

Declining Rates of Growth in Agricultural Output

In the early reform years, agriculture contributed heavily to the acceleration of economic growth following the disruptive years of the Cultural Revolution. In those years, the gradual introduction of market forces and de facto decollectivization resulted in substantial increases in output and value-added in agriculture. Decollectivization was key: by replacing the system of communal responsibility with one under which households became responsible for farming their own plots of land, the Chinese government created powerful incentives for Chinese peasants to increase agricultural output and reduce waste. As a consequence,

---

[6] Chow and Lin, 2002, pp. 507–530; and Wang and Yao, 2001.

[7] Chow and Lin, 2002, p. 528.

between 1980 and 1984, agricultural output rose at an average annual rate of 9.4 percent, while growth in value-added from agriculture rose at an average annual rate of close to 10 percent. According to the World Bank, from 30 to 60 percent of this increase was due to changes in incentives stemming from the introduction of the household responsibility system.[8]

By the late 1990s, growth in value-added in primary industry, predominantly agriculture, had tapered off to less than 3 percent per year, where it has since stayed. Much of the slowdown stemmed from a sharp deceleration in increases in crop output: in 2000, output of crops rose just 1.4 percent; increases in crop output averaged just 3.8 percent between 1996 and 2000. This slowdown stemmed both from resource constraints and from the consequences of government policies.[9] Almost all potential agricultural land in China is now tilled; in fact, environmental deterioration, the reversion of marginal cultivated lands to pasture or forest because of the introduction of more rational land use policies, and growth in industrial and urban areas is shrinking the total amount of land under cultivation. Although applications of fertilizers and crop protection agents could be increased, they are nearing economically optimal levels in many areas. Consequently, future increases in harvests are dependent on increasing yields through the more efficient use of such inputs as water, fertilizers, and plant protection agents and the introduction and dispersion of new seed varieties, cultivation techniques, and other technological advances, not greater application of agricultural inputs. Future increases in rural value-added will depend on shifting output toward more labor-intensive products and from the continued transfer of farm labor to other economic activities.

A return to more rapid growth in value-added from agriculture is likely to depend on institutional changes as well. China still lacks an efficient system of agricultural credit, clear land tenure arrangements with marketable land-use rights, and an efficient system of taxing agricultural activities. Rural markets for some agricultural products remain

---

[8] Nyberg and Rozelle, 1999, p. 4.

[9] Nyberg and Rozelle, 1999, Chapter 12.

distorted because the government manipulates purchase prices. For example, government policies directed at making China self-sufficient in grain production have led to some communes imposing physical production quotas for grains, constraining farmers from moving into more profitable crops. Liberalization of decisions on crop and product choice will be necessary to accelerate or even maintain growth in agricultural value-added.[10] However, even if these policy changes are made, agriculture is likely to be more a drag on future economic growth in China than a driver of that growth.

Widening Differences Between Rural and Urban Incomes

Not only has the slowdown in increases in value-added from agriculture placed a drag on overall economic growth, it has also exacerbated the difference in living standards between urban and rural areas. Although primary industry, of which agriculture is the predominant component, contributed just 14.5 percent of GDP in 2002, down from 31.2 percent at the beginning of the reform period in 1979, agriculture is still the largest single employer in China, accounting for 46.9 percent of total employment and 65.8 percent of total rural employment in 2000. Rural areas are still home to the bulk of the Chinese population: 63.8 percent of China's people live in the countryside.[11] Continued slow growth in this sector will have major implications for the livelihoods for close to half of China's population.

The ratio between urban and rural incomes in China is very wide, running 2.8 in 2000, up from 1.86 in 1985.[12] China now has one of the more inequitable rural/urban income distributions among developing countries. In terms of consumption, the difference between rural and urban households is more than twice as large in China as in India.[13] If growth in incomes in rural areas continues to lag that in urban areas, the difference in living standards between the countryside and cities will be exacerbated. Higher wages in urban areas have pulled

---

[10] Nyberg and Rozelle, 1999, p. 2.

[11] *China Statistical Yearbook,* 2001, pp. 91, 361.

[12] *China Statistical Yearbook,* 2001, p. 303.

[13] Nyberg and Rozelle, 1999, p. 12.

rural inhabitants to the cities, where many exist on the fringes of the urban economy. Because China has onerous laws concerning residency in urban areas; many rural migrants do not have legal permission to live in the urban areas where they work. Their tenuous legal status contributes to political unrest; the Chinese government has been focused on preventing economic dissatisfaction and unrest spilling over into political challenges to the leadership of the country. Economic and political resources are likely to continue to be directed at keeping unrest under control in the coming decades, thus competing for funds that might otherwise go to the Chinese military.

Financial Difficulties of the Town and Village Enterprises

Over the past two decades, improvements in living standards in the countryside have been increasingly tied to growth in economic activities other than agriculture. These activities have prevented the gap between urban and rural living standards from becoming even wider and have provided alternative sources of employment to rural residents other than migrating to the cities to search for work. Initially, most of these economic activities were generated by township and village enterprises (TVE), nominally municipally or collectively owned companies. TVEs engage in virtually the entire range of economic activities: construction, manufacturing, wholesaling, tourism, retailing, and transportation.

During the 1980s and early 1990s, TVEs were one of the most dynamic sectors in China's economy. In industry, output from TVEs increased at an average annual rate of almost 20 percent during the 1980s. However, this sector has fallen on hard times since 1996; output growth has slowed, and output from TVEs has even declined in some regions.

During this period, TVEs became the most important employer outside of agriculture in rural areas. In 1996, when employment peaked, these enterprises employed 27.5 percent of the rural labor force (Figure 2.4). Since 1996, employment has fallen, from 135 million to 128 million in 2000—in part because a number of TVEs have been privatized and in part because of bankruptcies and layoffs caused by increasing competition from private industry. Through 1999, growth in

Figure 2.4
Non-Agricultural Rural Employment in China

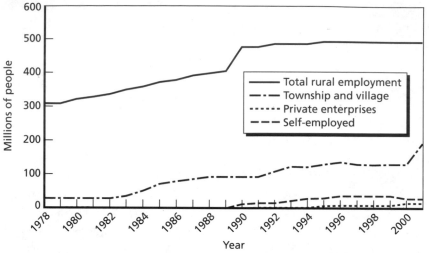

SOURCE: *China Statistical Yearbook*, 2001.
RAND *MG260-2.4*

employment in private businesses more than compensated for the fall in employment in TVEs. However, in 2000, employment outside of agriculture declined in rural areas as over 6,000,000 jobs were shed in that year.[14] The rural economy is suffering from increased competition from private businesses located in urban areas and from other Asian producers on export markets. Some of these problems are of their own making. Many TVEs have been poorly managed because local government authorities have often determined who is in charge. Appointments of managers have sometimes been made for political reasons rather than on the basis of competence. They also suffer from low levels of capitalization, overmanning, poor quality control, and process and product technologies that often lag those in privately run Chinese enterprises. In short, in the second half of the 1990s, rural China entered a period of slower economic growth as TVEs faced tougher competition. Although the private sector, including privatized TVEs, took over as the primary driver of employment growth, by 2000, in the country-

---

[14] *China Statistical Yearbook,* 2001, p. 111.

side growth in private-sector jobs was no longer compensating for job losses in TVEs and the state-owned sector. This period has yet to end. Until it does, the rural-urban economic divide in China will continue to widen.

The Service Sector: Current and Future Source of Employment

As growth in output and value-added from agriculture slowed, other sectors of the economy picked up the slack. Services are one such sector: Value-added from the services or "tertiary" sector rose at an average annual rate of 12.3 percent in the 1980s and 9.1 percent in the 1990s. As a consequence of this rapid growth, the importance of the service sector in China's economy has risen. In 1979, services contributed only 21.4 percent of GDP; by 2002, they accounted for 33.7 percent.[15] Services will remain a major driver of economic growth in the coming decades.

Despite this rapid growth, China's service sector has been relatively underdeveloped, a consequence of its centrally planned past. In middle-income developing countries, services account for more than half of GDP on average; in India and Indonesia, two lower-income developing countries, the service sector nevertheless contributes more than 40 percent to GDP compared with only one-third in China.[16]

Growth in value-added generated by services has been driven by increases in labor and capital as well as productivity improvements. Formal employment rose from 51,770,000 people in 1979 to 195,660,000 in 2000, i.e., by 378 percent, more than twice as fast as growth in the total labor force. In the second half of the 1990s, services were the only sector to show substantial growth in employment. Between 1994 and 2000, employment in services rose 26.6 percent, whereas employment in agriculture dropped 2.5 percent. Employment in industry rose 5.0 percent over the period but began to fall in 1999 and has continued to decline.

Although services tend to be a labor-intensive sector, in aggregate some branches, such as communications and transportation, are capi-

---

[15] *China Statistical Yearbook,* 2001, p. 50.

[16] World Bank, 1999, p. 64.

tal-intensive and absorb an appreciable amount of investment. Over the course of the reform period, the share of communications and transportation in total investment almost doubled, from 14.4 percent in 1980 to 28.0 percent in 2000, during a period in which overall investment rose an estimated 800 percent in constant price terms. Although the rise in output in transportation and communications has been impressive, up 7.4 times since 1980, the increase in investment has been much larger, up 21 times. Much of the increase in communications and transport output has been driven by increases in capital. In the future, services will remain the predominant source of new employment in China. However, the sector will also continue to absorb substantial shares of total investment. If the government needs to divert resources into other sectors such as health and education, growth in services will slow.

Will Industry Remain the Driver of Economic Growth?
Throughout the reform period, growth in industry value-added, especially in manufacturing, has been an important driver of overall growth: It averaged 11.5 percent per year between 1978 and 2001. However, this growth has been somewhat erratic. In the early 1990s, value-added in industry grew at superheated rates, averaging 17.7 percent between 1990 and 1995. The growth rate dropped to 9.0 percent per year after 1997. The rapid overall rates of growth in industry value-added have masked dramatic changes in the structure of industrial output. Over the past decade, the most rapid growth in output has been in the assembly and manufacture of household appliances and electronics. Taiwanese, U.S., and Japanese manufacturers have moved labor-intensive assembly operations to China. In addition, a group of indigenous companies are competing strongly in these sectors—in a number of cases, developing and exporting products that incorporate domestically developed technologies. Output of transport equipment and machinery has also grown strongly. Car production has risen from 35,000 units in 1990 to 607,000 in 2000 and over two million units in 2002. Production of intermediate goods, such as pulp, plastics, and construction materials like glass and cement, have also registered very rapid rates of growth.

Other industrial sectors have not fared as well. Increases in output of products from extractive and heavy industries have slowed dramatically in recent years; in some cases output has fallen. For example, the output of coal fell by 28.6 percent between 1996 and 2000 as the government cut subsidies and loans to loss-making coal mines. Rates of increase in output of pig iron, steel, crude oil, and bulk chemicals had fallen to the low single digits by 2000; output of timber and fertilizers also fell. Because the government placed tighter financial constraints on textile mills, cloth production fell sharply in 1996 as loss-making mills closed. Employment in textiles halved between 1995 and 2000; only in 2000 did the output of cloth return to its 1995 level. As a consequence, the shares of such traditional industries as coal mining, food processing, and textiles in total industrial output have fallen sharply, whereas those of electronics and telecommunications equipment, electrical engineering, and transport equipment have risen. By 2000, the last four branches of industry accounted for 21.7 percent of total industrial output; at the start of the reform period, they accounted for considerably less than 10 percent. The changes in the structure of industry reflect the effect of market pressures on traditional manufacturers.

The increased manufacturing output and the shift in structure have been driven in part by changes in the control of enterprises and by new entrants. As of 2000, the share of industrial output accounted for by non-state-owned enterprises was 52.7 percent; at the start of the reform period the share of industrial output generated by non-state-owned enterprises was almost negligible. Enterprises controlled or involving foreign ownership also play a major role in Chinese industry: In total, "foreign-controlled" enterprises produced 27.3 percent of Chinese industrial output in 2000, of which 12.3 percent was produced by enterprises in which entrepreneurs from Hong Kong and Macao held controlling or substantial stakes and 15.0 percent was produced by enterprises in which foreign investors held a substantial stake. Before 1978, there were no such enterprises in China.

In the future, growth in industrial output is destined to slow. China has already undergone wrenching changes in employment and output in traditional industries. These will continue as state support for companies continues to fall, China's economy opens up to more import competition, and less-efficient firms fall by the wayside.

Foreign Direct Investment: A Diminishing Driver of Growth

Early on, attracting foreign direct investment (FDI) became an important part of China's economic reforms. In July 1979, the Law on Sino-Foreign Equity Joint Ventures was promulgated, which set up the legal framework for foreign direct investment by permitting foreign investors to create joint ventures with Chinese enterprises.[17] Before the passage of this law, FDI had not been permitted. Initially, foreign investors were hesitant, and most investment came from the Chinese diaspora in Hong Kong and other Asian countries. In 1982, the first year for which we were able to obtain data, FDI ran $430 million, less than 0.2 percent of Chinese GDP at market exchange rates in that year.

The Chinese government has made a number of regulatory and legal changes since the 1979 law to attract foreign investors, including the creation of various types of special economic zones in which foreign investors enjoy special privileges. Depending on the sector, foreign investors have been exempted from paying corporate income taxes for two years after the first year of profitable operation and have been given another three years in which their corporate income taxes are reduced by 50 percent.[18] This favorable treatment has contributed to attracting steadily rising inflows of FDI in the 1980s. However, even in 1990 FDI ran just $3.5 billion, less than 1 percent of GDP. It was only after that date as the legal environment became more benign that FDI really took off, rising to $44.2 billion in 1997, 4.9 percent of GDP (Figure 2.5). Cumulatively, FDI ran $430 billion by 2002, over one third of China's GDP at market exchange rates.[19]

These inflows of FDI played an important role in sustaining Chinese economic growth rates in the 1990s. They not only provided

---

[17] Liu, 2002, p. 581.

[18] Liu, 2002, p. 579.

[19] International Monetary Fund, various years. The IMF data are balance-of-payments data and therefore include actual inflows. If reinvested earnings by foreign investors in China were included in these data, the numbers would be higher.

Figure 2.5
Inflows of Foreign Direct Investment into China

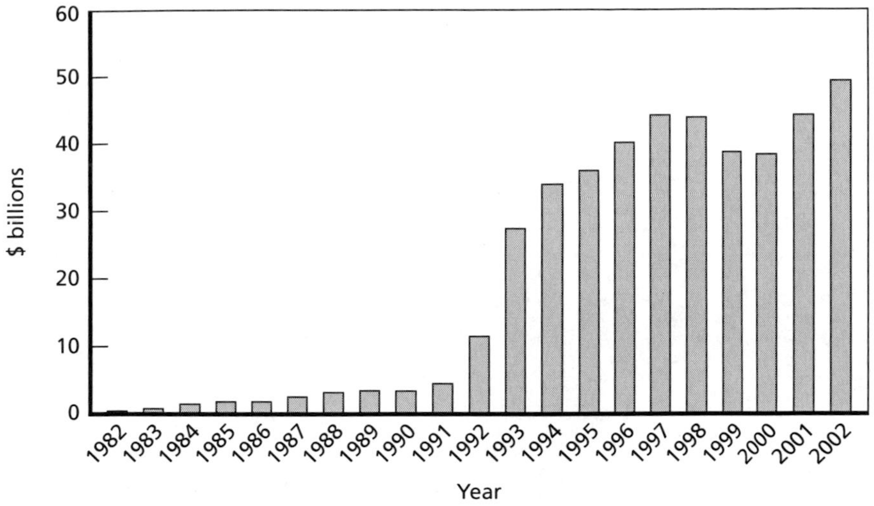

SOURCE: IMF, International Financial Statistics, various years.
RAND *MG260-2.5*

China with capital but also generated associated inflows of technologies, new managerial and organizational techniques, better accounting practices, and improved access to foreign markets. In addition, they generated spillover effects by spurring increases in productivity in competing and associated Chinese-owned firms in manufacturing.[20] In fact, these contributions to improving productivity have been more important than the inflows of capital per se. On a cumulative basis, gross domestic savings have exceeded domestic investment throughout the reform period: In other words, China has been a net exporter of capital.

Inflows of FDI into China can be separated into two major streams: investments from ethnic Chinese located outside of China, including Hong Kong and Macao, and inflows from the developed market economies of the European Union, Japan, and the United States. The two flows have quite different characteristics and eco-

---

[20] Liu, 2002, p. 590.

nomic effects. Inflows of FDI from the Chinese diaspora account for the bulk of FDI in China: Hong Kong alone accounts for over half the total. Taiwan is an increasingly important source of capital as well, although it is difficult to track this investment because it is channeled through third countries like the British Virgin Islands or the Cayman Islands for legal reasons. Inflows of foreign investment into China from the diaspora tend to be concentrated in labor-intensive manufacturing and assembly operations. Many of these operations are focused on exporting products abroad, especially to developed market economies. These operations have had a major impact on growth in manufacturing output, especially in such consumer items as toys, clothing, shoes, and electronics. In many instances, they have displaced similar operations in Hong Kong and Taiwan, where labor costs are now substantially higher. As overseas Chinese have gained confidence in the business environment in China, these flows have risen. However, a sharp deterioration in the climate for foreign investment would likely result in a sharp decline.

Part of this "foreign" investment from the Chinese diaspora, as much as 25 percent, involves reimports of capital by mainland Chinese.[21] Like Russia, China has been an exporter of capital in recent years. Because of legal restrictions and the desire to shield income and wealth from the Chinese tax authorities, Chinese investors find it beneficial to export capital. Although some of this capital is invested abroad, an appreciable share is reimported into China as "foreign" investment. In this manner, mainland Chinese can take advantage of the favorable tax and regulatory treatment the Chinese government provides foreign investors.

Foreign investors from developed market economies have been more focused on selling into the domestic Chinese market. They have invested in domestic distribution systems and the manufacture of more technologically sophisticated products, such as automobiles and microchips, than have investors from the Chinese diaspora. Consequently, Japanese and U.S. investors have located their operations in areas

---

[21] Harrold and Lall, 1993.

where the labor force is relatively more educated. In contrast, overseas Chinese are attracted to areas where labor costs are lowest, as long as the transportation systems are adequate for exporting products.[22]

Although China remains an attractive market, in recent years, inflows of FDI into China have been higher than into any other country in the world, including the United States. Continued increases in FDI are unlikely. In fact, a more likely scenario would be a decline in FDI in the coming years once large multinational corporations have built up their sales and service networks in China. As this process slows, FDI may fall, which is likely to retard China's economic growth.[23]

Exports Will Become Less of a Factor in Growth

A key thrust of the Chinese government's reform policies has been to open China's economy to the outside world. During the Mao era, China was very closed. A large part of its trade took place with the former Soviet bloc, in part because of the U.S. and Taiwanese embargoes but primarily because of Mao's distrust of trading with the "enemy." In general, Mao preferred not to trade at all but to try to make China self-sufficient. During this period, the Chinese economy operated according to the dictates of central planning. The state exercised a monopoly on foreign trade; all foreign transactions were undertaken through state-owned foreign trade companies. Domestic prices were divorced from prices on international markets. As a consequence, enterprises had little incentive to export and were protected from foreign competition. Not surprisingly, China played a minor role in world trade, accounting for just 0.81 percent of the total in 1978, less than the shares of many smaller European economies. Exports were only 4.6 percent of GDP.

As China opened to the outside world during the 1980s and 1990s, its comparative advantage in labor-intensive goods resulted in strong export growth. Exports grew at an average annual rate of 15.7 percent in current dollar terms between 1978 and 2002. These exports are largely

---

[22] Fung, Iizaka, and Parker, 2002, p. 576.

[23] In a study published in 2003, RAND assessed the potential impact on the Chinese economy of a reduction in foreign direct investment. The authors argued that a $10 billion reduction in inflows of FDI might reduce average GDP growth rates by 1.6 percentage points. See Wolf, Yeh, Zycher, Eberstadt, and Lee, 2003, pp. 154–156.

manufactured goods assembled from components or materials imported from abroad. A substantial share is produced by enterprises with some foreign ownership. These enterprises generated 48 percent of total Chinese exports in 2000. Like companies that have located assembly plants on the U.S.-Mexican border, these enterprises are primarily interested in utilizing China's low-cost workers for labor-intensive operations. The composition of China's exports and imports reflect the key role played by these assembly operations. Textiles, clothing, shoes, leather products, and headgear accounted for 27.6 percent of China's total exports in 2000. Electrical equipment and electronics accounted for an additional 21.8 percent. In other words, labor-intensive products and assembly work accounted for half of total exports in 2000.

The Chinese government permits companies that assemble products for export to import components duty-free. Not surprisingly, these export operations import substantial quantities of components. Imports by enterprises with foreign ownership have run well over 90 percent of exports from these companies, reflecting the large share of imported components in exports. This figure also reflects "leakage" because these companies "redirect" products from exports to the domestic market, often illegally.

The other half of China's exports consists of a variety of goods, many of which are produced by state-owned enterprises. China exports substantial quantities of bulk chemicals, crude oil, metals, and metal products. Most of these products are capital-intensive; many are only marginally profitable. These exports do not appear to play to China's comparative advantage in labor-intensive manufacturing. In some cases, they are the result of past centrally dictated investments or overinvestment by enterprises in particular sectors, which have left China with excess capacity that enterprises utilize to export. Once these facilities have been constructed and are up and running, the cost of producing additional output is relatively small, so enterprises prefer to run close to capacity and export—even if export prices are relatively unattractive—rather than operate at lower rates of capacity utilization. Higher operating rates also make it possible to keep more workers employed, thereby satisfying local and national government desires to avoid layoffs.

Chinese enterprises also produce a variety of traditional products for export. Exports of prepared fish, canned fruits and vegetables, dyes, and perfumes often are either labor-intensive or rely on locally produced materials. These products tend to have the highest share of value-added among Chinese exports.

In the future, China's exports will undergo some massive shifts. First, continued double-digit growth in exports will become much more difficult, if not infeasible, because China already holds such large shares of global markets. Between 1995 and 2002, the average rate of growth in exports fell to 11.8 percent from 19.1 percent between 1990 and 1995. Although export growth has surged again in 2003 and 2004, it is likely to slow again because of exchange rate pressures as well as market saturation. Because the Chinese economy has grown while other economies in the region have faltered, exchange rates of competitor nations have depreciated sharply while China's government held the renminbi steady against the U.S. dollar. Large inflows of foreign direct investment, the strength of China's external balances, and U.S. pressure may well result in the appreciation of the renminbi in the near future. An appreciating renminbi will put increasing competitive pressure on Chinese producers of such labor-intensive products as clothing and shoes from suppliers in lower-wage countries like Bangladesh or even Indonesia.

Some Chinese exports of raw materials and intermediate goods are likely to disappear. China has already become a major importer of oil (a commodity it once exported) and steel. Overcapacity in other capital-intensive industries is either disappearing because of rising domestic demand or will be reduced as older operations that use less-efficient technologies are shut down, especially if the Chinese government maintains its current policies of imposing financial discipline on state-owned enterprises and banks. In short, increases in exports, a key driver of economic growth over the past two decades, are destined to slow in the coming decades.

Balance-of-Payments Pressures Are Unlikely to Constrain Economic Growth

Many developing-country economies run aground because of balance-of-payments crises. They import capital during periods of rapid growth. These inflows, which show up in the form of surpluses on their capital and financial accounts, are mirrored in current account deficits as capital inflows make it possible for a country to import more than it exports. However, if their economies overheat or experience an external shock such as a rise in energy prices or global financial market interest rates, these inflows dry up and the country has to adjust.

China has had a strong record of maintaining external balance. Although it ran current account deficits in the mid-1980s, it has run current account surpluses for all but one year since 1989, despite substantial inflows of foreign direct investment. As a consequence, China has built up large foreign reserves, primarily in dollars. Official international reserves have risen from $2.5 billion in 1980 to over $370 billion in 2003.

China's current account surpluses are somewhat exaggerated, however. As noted above, Chinese investors have been taking advantage of the Chinese government's tax and regulatory concessions to foreign investors to ship capital abroad and then reimport it as "foreign investment." Despite capital controls exercised by the People's Bank of China, capital exports are possible through under-invoicing exports or over-invoicing imports. The amount of these capital exports has declined in recent years as the Chinese government has begun to offer more equal treatment to domestic enterprises.

For the immediate future, China is likely to continue to register modest current account surpluses on the order of up to 2 percent of GDP. Chinese households, like their Japanese counterparts, have a propensity toward high rates of saving. But because China's population is aging rapidly, these current account surpluses are likely to shrink because savings rates are likely to fall as the elderly draw on savings for retirement. Savings rates are also likely to fall as China becomes more of a "consumer society." As incomes rise and consumers become more confident about their future economic outlook, consumption and expenditure patterns shift. Consumers begin to borrow to purchase housing and consumer durables, thus reducing savings rates. Changes in tastes and incomes make imports more affordable and desirable. Imports of consumer goods have already been increasing rapidly because China has cut tariffs and liberalized import procedures. China's commitments under the World Trade Organization (WTO) agreement, coupled with strong pressure to open its markets exerted by the European Union (EU) and the United States suggest that trade liberalization will continue in the coming decade, contributing to continued strong growth in imports. Although a balance-of-payments crisis is highly unlikely in this decade because of China's high domestic savings rates and the government's conservative management of its external accounts, external imbalances could become a problem in the medium term.

Threats to Growth from the Financial System

The condition of China's state-owned banks poses one of the greatest threats to sustained growth. Although rates of return on private investment in China appear to be substantial, the severe credit-quality problems of China's major banks attest to the poor use of borrowed funds by state-owned enterprises. According to the International Monetary Fund (IMF), the state-owned banks' nonperforming loans run between 25 percent and 75 percent of total bank assets.[24] Taking another approach, the IMF estimates that bad loans could run 50 to 75 percent of GDP.[25] These bad loans are a major problem. First, they translate into very low average rates of return on invested capital because so much investment has had to be written off. Second, the Chinese government will have to recapitalize state-owned banks when the banks write down these bad loans. Bankruptcy of the state-owned banks is not an option because they comprise such a large share of the banking sector.

How did China's financial system come to such a pass? During the course of the reform, enterprises were supposed to become self-sufficient. The government withdrew price controls and subsidies with the intent of forcing state-owned enterprises to generate sufficient cash flow to fund their operating costs. Bank finance replaced government grants for investment. As market forces became stronger and state-owned enterprises faced competition for the first time, many enterprises proved unprofitable. They turned to the state-owned banks to finance ongoing operations as well as investment. The Chinese government, especially the finance ministry, faced its own financial constraints. The government chose to turn a blind eye to the poor financial state of the enterprises. In many instances, the government encouraged or even

---

[24] IMF, 2002, Chapter One, Box 1.4.

[25] IMF, 2002, Chapter One, Box 1.4.

forced the state-owned banks to advance loans to those enterprises. In effect, the government substituted bank credits for subsidies.

The state-owned banks were complicit in this process. Until 1979, the People's Bank of China (PBC) had handled virtually all financial transactions in China. In that year, the Agricultural Bank of China and the Bank of China were created from departments in the PBC. In 1984, the Industrial and Commercial Bank of China was created from the remaining commercial operations of the PBC and the PBC became the central bank. In 1985, the People's Bank of China was hived off from the Ministry of Finance.[26] These four large state-owned banks lacked the analytical capability, operational freedom, and commercial incentives to properly assess credit risk. Bank loans were determined by local managers, often in conjunction with the local government and party leadership and frequently with the client, and they were often made for political or personal reasons. Credit quality has suffered as a consequence. Not surprisingly, the state-owned and communally owned enterprises that have borrowed heavily from the state-owned banks now find that they are unable to repay these loans because their investments failed to pan out or the borrowed funds were used to finance loss-making operations rather than to invest in new capacity.

The banks' problems have been managed by transferring bad loans to asset management companies. The asset management companies have been kept afloat by central bank guarantees. The Chinese government is now being forced to recognize these bad debts; their cost has moved back to the government budget. Recapitalizing the state-owned banks to stave off a collapse of the financial system will result in an increase in government debt. Assuming that these loans will increase government debt by 50 percent of GDP, the cost of servicing these debts will be on the order of 2 to 3 percent of GDP annually, assuming an average interest rate on government paper of 4 to 6 percent. Unfortunately, despite the attempts to address the bad debt problem, the large state-owned banks still seem to lack the capacity to make loan decisions on the basis of objective measures of creditworthiness. Consequently, the problem of poor allocation of investment endures, reducing the

---

[26] Holz, 2000, p. 87.

productivity of investment and continuing to store up future problems with credit quality and the banking system.

The Impact of Demographic Changes on the Chinese Economy

Urbanization and Internal Migration

China has been undergoing a dramatic transition from a largely rural, agrarian country to a more urbanized industrial economy. Large numbers of rural inhabitants have migrated to cities and formerly rural areas have been incorporated into urban areas as cities have engulfed smaller towns and villages. Reflecting these changes, the share of the population living in urban areas rose from 17.9 percent in 1978 to 36.2 percent in 2000. This process will continue. Extrapolating from past trends, over half of China's population will live in urban areas by 2020, by which time the urban population is likely to run 758 million, more than 4 times the 172.5 million urban dwellers in 1978.

People migrate not only from country to town in China but from region to region. A substantial number of people have migrated from west to east as the eastern coastal provinces have drawn in labor—primarily young men, but increasingly young women—from the poorer central and western provinces to staff factories and construction sites. There has also been movement from the northeastern provinces to the south as heavy industries in the northeast have downsized.

The ongoing urbanization of China has a number of ramifications for government spending and budgetary expenditures. The Chinese government has been spending heavily on infrastructure in recent years, and sewage and water systems have been constructed for the burgeoning cities. These expenditures will need to increase in the coming decades to create the urban infrastructure needed to house this massive shift in population.

Migration also increases the demand for personal transport as migrants return home to visit friends and family. These visits have already had a massive effect on personal transportation in China: Total passenger kilometers jumped 7.5 times between 1978 and 2001. The resulting strain on railroads, roads, and air transport has been driving massive investments in the transport sector.

The Aging of China

China is the most populous country in the world, with 1.28 billion inhabitants in 2002. It is also home to one of the world's most rapidly aging populations. In part because of China's one-child policy and in part because of increasing urbanization and rising incomes, fertility rates have plummeted. Between 1960 and 1978, the beginning of the reform period, population growth rates averaged 2.2 percent per year. As of 2002, the estimated average number of children borne per woman had fallen to 1.69, substantially below the replacement rate. Despite these low fertility rates, rising life expectancy and the large number of women of childbearing age have led to continued population growth, although population growth rates are slowing. The rate of population growth dropped to just 0.9 percent in the first years of the 21st century. After 2037, continued low fertility rates will lead to an absolute fall in China's population.

Meanwhile, average life expectancy has been rising, from about 65 years in 1981 to 71 in 2001.[27] Rising life expectancy and declining fertility rates have resulted in a very sharp increase in the share of the elderly in China's total population. Figure 2.6 compares the proportion of the total population in China that is over 60 with similar data and projections for Japan and the United States. Although China's population is still considerably younger than that of either Japan or the United States, it is aging much more rapidly. In 1990, only 8.5 percent of China's population was over 60; by 2030, that percentage is projected to rise to 24.0 percent. Between 1990 and 2030, the share of the elderly in China's population is projected to almost triple; in the United States, that share is forecast to rise by half.

The recent aging of China's population has created a demographic "golden period." In the 1960s when population growth was running 2.35 percent per year, a substantial share of the population consisted

---

[27] U.S. Census, International Data Base, http://www.census.gov/ipc/www/idbacc.html; and the World Bank, World Development Indicators data base, http://www.worldbank.org/data/countrydata/countrydata.html.

Figure 2.6
The Share of People over 60 Years of Age in the Populations of China, Japan, and the United States

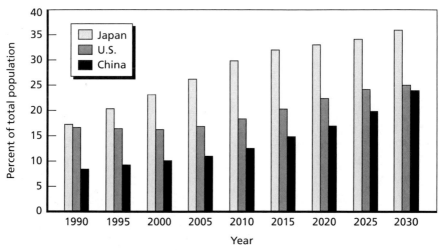

SOURCE: U.S. Bureau of the Census, International Data Base, http://www.census.gov/ipc/www/idbacc.html.
RAND *MG260-2.6*

of dependents, mostly children. Even as late as 1982, 38.6 percent of the population was not of working age, i.e., younger than 15 years old or older than 60. By 2002, as fertility rates fell, the share of the population accounted for by dependents had shrunk to 31.1 percent. Conversely, the share of China's total population of working age had risen to 68.9 percent. The influx of these individuals into China's labor force has contributed to recent rapid rates of economic growth. This golden period will hit its apex in 2010, when 72.1 percent of China's total population will be of working age—2.6 people of working age for every dependent, compared to the current ratio of two to one. After 2010, the golden age begins to fade as the number of elderly rises and the number of new entrants into the workforce falls. By 2050, there will be only 1.5 people of working age for every retiree or child in China. As the potential labor force stagnates and then falls, the number of dependents will begin to rise, imposing social and fiscal burdens on China's citizens and government. Thus, from 2010 onward, the Chinese government will begin to face increasing demands for spending on pensioners as it faces the same fiscal and social challenges associated with an aging population that already confront governments in Europe and Japan.

Aging and the Military

As China ages, the number of young men between 18 and 22 will fall sharply, from 65.7 million men in 1990 to 49.7 million in 2025 (Table 2.1). As the table shows, there is a substantial amount of variation in the size of draft age cohorts because of past booms and busts in births. The draft pool plummeted between 1990 and 2000 but will recover sharply by 2010 as past trends in births echo through the age structure. Superficially, despite the general decline in the pool of draft age men, it should not be difficult for the PLA to meet its recruiting goals. The Chinese armed forces currently comprise 2.25 million people with another 500,000 to 600,000 in the reserves. Moreover, the PLA has been downsizing its forces, slowly shifting from a large, conscript-based land army to a smaller, more modern force. The PLA will need to take at most 3 percent of the total pool to meet staffing levels. With a pool of this size, the PLA should have little difficulty in finding enough recruits to staff a shrinking armed force, provided pay and career prospects remain attractive.

The key here will be providing attractive alternatives. The PLA has drawn heavily from rural areas because, even at the very low wages of the past, a career in the army was more remunerative than working the land. However, as more and more young men grow up in urban environments or have opportunities for work in nonagricultural occupations, the PLA will need to raise wages to attract enough soldiers. As noted in Chapter Four, the PLA has already had to raise wages sharply to keep its troops happy. In addition, if the PLA is to modernize successfully, it will need to attract more highly educated soldiers with technical skills. These individuals have a much broader range of economic opportunities than uneducated rural laborers do. The PLA will have to provide them with competitive wages and career opportunities, if it is to attract individuals of the caliber that it needs.

Table 2.1
Pool of Draft-Age Chinese Males

Year	Number (millions)
1990	65.75
1995	61.51
2000	49.04
2005	50.58
2010	61.73
2015	53.19
2020	50.01
2025	49.74

Future Economic Growth

We conclude the chapter with a discussion of prospects for future economic growth in China. The purpose of this section is to quantify the likely future size of China's economy to provide a basis for projecting future levels of resources the country may have to draw upon for military spending. We first review the accuracy of past numbers concerning rates of economic growth in China and then discuss the main drivers of future growth and how they are likely to develop through 2025. We conclude by presenting various forecasts of the size of the Chinese economy in 2025 and selecting a base-case forecast that we will use for the projections of military spending developed in Chapter Seven.

Biases in Chinese Statistics on Economic Growth
Gross domestic product is a construct for measuring economic output. It suffers from the problems of all such constructs. First, it excludes some important economic activities. Household tasks, for example, are not included in GDP unless someone is paid to perform them. As consumers become wealthier, they purchase more restaurant meals or pay for cleaning and other services that they formerly provided for themselves in the home. The substitution of paid services for household activities results in an increase in reported GDP, even though there has been no increase in output. This shift from households providing

their own services to purchases from the market may have a significant impact on reported GDP.

A second problem stems from the choice of prices to be used in measuring changes in economic output. Relative prices of output in rapidly growing sectors tend to fall over time. For example, the real price of computers, i.e., the cost of buying computers of the same performance, has fallen dramatically over the past three decades as a result of technological improvements. If one measures the value of computers over time by keeping prices constant, the recent rapid growth of the computer and information technology industries would have a much larger effect on GDP if prices at the beginning of the period are used rather than prices at the end. Because changes in economic structure and output in China have been so great, Chinese growth statistics are very sensitive to these changes in relative prices. In the case of China, the use of prices from earlier years to evaluate current changes in economic output tends to bias growth rates upward.

A third problem is inadvertent or advertent biases in calculating GDP on the part of China's central statistical office. A number of institutions and scholars have attempted to remeasure past growth in Chinese GDP.[28] As shown, in Table 2.2, their recalculations of Chinese GDP growth rates reduce average annual growth rates by anywhere from 0.9 to as much as 4.9 percentage points.

Of the figures in Table 2.2, those published by Professor Thomas G. Rawski of the University of Pittsburgh show the largest divergence from official growth rates. Rawski argues that the Chinese statistical authorities have adopted a deliberate policy of exaggerating figures on GDP growth since the Asian crisis of 1997. He notes that since 1998, during a period when official figures show continued rapid economic growth, energy consumption has fallen, growth in employment has slowed to about 1 percent per year, and rural income has declined. Provincial data also began to diverge sharply from the national aggregates during this period.[29] Rawski argues that under pressure from the Chinese Communist Party leadership to paint a cheerier picture of the

---

[28] Taken from Lardy, 2003, p. 12, and Rawski, 2001.

[29] Rawski, 2001.

Table 2.2
Alternative Measures of Rates of Growth in China's GDP

Institution	Period	Official Rate (Percent)	Reestimated (Percent)	Difference
World Bank	1978–1995	9.4	8.2	1.2
World Bank	1978–1986	9.7	8.8	0.9
World Bank	1986–1995	9.2	7.9	1.3
OECD	1986–1994	9.8	6	3.8
Rawski	1998–2001	7.7	2.8	4.9
China National Economic Research Center	1978–1998	9.7	8.4	1.3

SOURCE: Lardy, 2002, p. 12, and Rawski, 2001.

economy, the Chinese statistical authorities have been inflating Chinese statistics on economic growth. He quotes a number of Chinese economists and outside observers to support this contention.

On the other hand, some factors serve to underestimate growth in China. In particular, Chinese entrepreneurs work hard to hide revenues from the tax authorities. These efforts extend to the statistical authorities as well. To the extent the statistical authorities fail to capture this activity, the national estimates of GDP are underreported. Failure to capture unreported or "informal" economic activity has implications for estimates of economic growth as well as for the size of the economy. Most unreported economic activity is in the private sector. Because the private sector has been growing more rapidly than the state sector, if the statistical authorities do a worse job of capturing increased economic output in the rapidly growing private sector than in the slower growing state sector, estimates of economic growth will be biased downward.

Prospects for Growth Through 2025

Key Factors Affecting Future Growth in China
Whatever the true story concerning the precision of GDP growth rates in China, all statistical series show that growth slowed in the second half of the 1990s. Official figures on growth in GDP fell from an aver-

age annual rate of 12.1 percent between 1991 and 1996 to 7.6 percent between 1997 and 2002. As noted above, the slowdown in growth has been driven by a variety of factors, including a fall in the rate of growth of agricultural output and the decline of heavy industry. However, a number of factors—from increases in foreign direct investment to growth in manufacturing and services—have continued to drive increases in GDP. Of these factors, we believe the following will be the most important determinants of future economic growth:

1. Agriculture will remain a drag on growth. Arable land is shrinking, the sector remains the residual employer, and the easy productivity boosts provided by the introduction of markets and applications of more agricultural inputs such as fertilizers have already been achieved. Government policies aimed at keeping China self-sufficient in grain will also retard growth in this sector because they will slow the shift from grain to more-profitable crops.

2. Export growth will slow. China already accounts for a substantial share of global trade, making rapid export growth more difficult. A slowdown in export growth will result in slower growth in manufacturing, which was the driver of China's economy in the 1990s. Moreover, by many measures, China's trade regime is already fairly liberal. Consequently, the contribution to economic growth provided by foreign trade is likely to be smaller in the coming years than it was in the recent past.

3. Weaknesses in the financial sector are likely to retard growth, at least until the financial sector is restructured. China has not yet created an effective financial system for efficiently allocating capital. Although private equity investment, foreign and especially domestic, appears to have generated substantial rates of return, China's state-owned banks continue to make ill-advised lending decisions. Although the Chinese government is taking steps to recapitalize banks, until capital markets are liberalized and the state-owned sector faces more competition from private competitors, especially foreign banks, China's financial sector will probably continue to channel savings into projects or enterprises that fail to repay their loans. This has been the pattern in other de-

veloping countries. Moreover, once the monetary authorities in developing countries are forced to recognize the consequences of bad loans, those countries have invariably experienced a period of slower growth as banks are recapitalized and the operations of bankrupt firms wind down because of cutoffs in bank finance.

4. GDP growth will slow as the labor force stagnates and then shrinks. After 2005, the rate of growth in the working age population will begin to decelerate; by 2017, the working age population will begin to decline. The reduction in the size of the labor force will slow the rate of growth in GDP by at least a percentage point.

5. China's rate of domestic savings will fall, curbing increases in investment. The aging of China's population will lead to lower savings rates as the elderly use savings to provide for their retirement and the government diverts more of the budget from public investment to pension and health payments. The expansion of consumer credit and the greater availability of consumer goods appear to be reducing the propensity to save, at least among better-off urban consumers. As savings decline, the share of investment in GDP will almost certainly fall.

The confluence of these factors is likely to lead to deceleration of growth rates of GDP. The retarding effect of agriculture on growth, the weaknesses of China's financial sector in terms of effectively allocating capital, the slowdown in the rate of growth of the labor force, and the shift of final output from investment to consumption as the population ages make it difficult to argue that China will maintain the growth rates of the past two decades. Historical precedent also suggests that growth will slow. Growth has slowed in the once fast-growing Asian economies of Japan, Singapore, South Korea, and Taiwan. One can argue that China's per-capita GDP ($4,227 at purchasing power parity exchange rates and $963 at market exchange rates in 2002) is still so low that a number of years of rapid growth remain in store for the country. However, by the end of this decade, the factors cited above are likely to work to slow rates of growth in GDP as they have in other Asian countries.

Projections of Growth in GDP

A number of institutions and individuals have engaged in forecasting long-run economic growth rates for China, including RAND (Table 2.3). Rather than add to the cacophony of forecasts of long-run growth rates, we have chosen to review some of those forecasts and select one among them rather than develop our own independent forecast for this study. We believe that we would have been unlikely to generate a substantially more accurate forecast had we chosen to generate our own rather than those that have already been developed.

Because of the multitude of variables that impinge on economic growth, none of the forecasts will be precisely right. The world is too uncertain and too changeable for that to be possible. But the forecasts do help to quantify the influence of overall trends in a few key drivers of economic growth by providing a systematic, quantitative framework within which to evaluate likely future trends in China's economy. For our purposes, these forecasts provide an internally consistent quantitative basis for projecting the resources that are likely to be available for military spending in the years ahead.

Because of the many factors that will serve to slow growth, we have chosen to use the forecast of 5 percent average annual growth through 2025 that was generated by past RAND work on China.[30] We

Table 2.3
Forecasts of Average Annual Long-Term Growth Rates for China

Source	Period	Rate
U.S. Department of Energy[a]	1999–2020 base case	7.0
U.S. Department of Energy[b]	1999–2020 low-growth case	3.9
World Bank[c]	2002–2020	6.1
RAND[d]	2002–2015	5.0

[a]Energy Information Agency, 2002, p. 182.
[b]Energy Information Agency, 2002, p. 222.
[c]The World Bank, 1997.
[d]Wolf et al., 2000, p. 14.

---

[30] Wolf, Bamezai, Yeh, and Zycher, 2000, p. 14.

do not project an immediate slowdown. Rather, we project growth of 7 percent per year through 2010, gradually declining to 3 percent per year in 2025. These rates of growth are substantially lower than the average annual rate of growth of 8.7 percent reported for the past quarter century. However, all the forecasts project a slowdown in growth. They range from an average annual rate of 3.9 percent in the low growth scenario of the U.S. Department of Energy to 7.0 percent for the reference case scenario for the same institution. The World Bank forecasts average annual growth rates of 6.1 percent for the 2002–2020 period. We chose the RAND forecast because if falls in the middle of the other forecasts. It also better incorporates the many factors that will serve to slow growth in China's GDP in the coming years. It is also more consistent with historical trends in GDP growth in other Asian economies. In all the Asian countries that enjoyed very rapid rates of growth these countries, long-run average growth rates have decelerated, in some cases, like Japan, very dramatically. These slowdowns and in some cases recessions frequently occur following three decades of rapid growth. Although growth resumes after these slowdowns or recessions, it does not return to past rates.

What does 5 percent average annual growth imply concerning the future size of China's economy? At that rate of growth, China's economy would triple between 2003 and 2025 in constant prices. As China's population is projected to grow just 14 percent over this same period, per-capita GDP is projected to rise 260 percent. In other words, the average Chinese citizen will generate 2.6 times more output in 2025 than he does today.

What do these growth rates mean in terms of the size of China's economy compared to that of the United States? If these growth rates were to be applied to China's GDP as measured by purchasing power parity exchange rates, the Chinese economy would run 17.3 trillion 2001 dollars in 2025. Assuming the United States grows at an average annual rate of 3.0 percent over the same period, the U.S. economy would double by 2025. In other words, in 2025 China's economy would be 88 percent of projected U.S. GDP in that year, but 73 percent more than the U.S. economy in 2002 at current PPP exchange rates (Table 2.4). On a per-capita basis, this outcome is not quite so

Table 2.4
RAND Forecasts of China's GDP

Type of Exchange Rate	GDP in 2002	GDP in 2025	Share of Projected U.S. GDP in 2025	Share of Projected Per Capita U.S. GDP in 2025
	Trillions of 2001 dollars	Trillions of 2001 dollars	Percent	Percent
PPP exchange rate	5.43	17.31	88	20
Market exchange rate	1.24	3.95	20	5
Projected combination of PPP and market exchange rates	3.54	9.45	48	11

stunning. Under that scenario, China's per-capita GDP is projected to run one-fifth of projected per-capita U.S. GDP in 2025 and one-third of current per-capita GDP.

The use of purchasing power parity exchange rates provides an inflated picture of the future size of China's economy in comparison with that of the United States. The current PPP exchange rates are substantially greater than market exchange rates because of the relatively lower cost of basic consumer goods and services in China. As China grows, the importance of these items in the typical market basket of Chinese consumers will fall while the relative cost of these items will rise because of increases in real wages. Consequently, PPP exchange rates will not remain static; the gap between the commercial or market exchange rate and the purchasing power rate will narrow by 2025.

As noted above, estimates of GDP at PPP exchange rates are useful for comparing living standards but often do a poor job of reflecting the purchasing power of defense budgets, especially for procurement. Advanced weapon systems incorporate technologically sophisticated materials and components that are sold at world market prices; the relatively lower cost of housing, personal services, and food that are captured using PPP exchange rates have little relevance for expenditures on these items. Because of the expected changes in the PPP exchange rate and because of the weaknesses of estimates of PPP rates for measuring military spending, GDP measured at market exchange rates offers an alternative measure of the resources available for spending on defense. This measure is preferable for estimating resources available for procurement of military equipment.

To forecast China's GDP at market exchange rates, GDP is converted to dollars at market exchange rates for a base year (2001 in this case). Projections of GDP in renminbi are converted using this exchange rate throughout the forecast period under the assumption that the renminbi/dollar exchange rate will remain constant in real terms over the forecast period. Using this method, China's GDP would run $3.9 trillion in 2001 dollars in 2025, 39 percent of current U.S. GDP and 20 percent of projected GDP, assuming the United States grows at an average annual rate of 3.0 percent over the same period (Table 2.4). Forecasts of China's GDP converted to dollars using the current market exchange rate are only a fraction of forecasts converted using the current PPP exchange rate.

The assumption that the current market exchange rate will remain constant in real terms over the forecast period is not very credible. Over the course of the next two decades, the market exchange rate should strengthen. Rapid rates of growth in productivity will enable China to produce a given quantity of exports with fewer inputs, resulting in an improvement in China's terms of trade. If China's real effective exchange rate were to remain constant, exports would rise more rapidly than imports, leading to an expanding trade surplus. Foreign exchange market pressures stemming from such a trade surplus would then serve to push up the real effective value of the currency.

Because both the market and PPP renminbi/dollar exchange rates are almost certain to change over the course of the coming two decades, we have adopted a third method to project China's GDP in 2001 dollars. First, we project changes in the PPP and market exchange rates in terms of 2001 dollars over the course of the next two decades. PPP exchange rates were projected by forecasting changes in relative per-capita GDP (income) between the United States and China. As noted above, the large gap between market and PPP exchange rates in China is in great part due to China's relatively low wages compared to those in the United States. As China becomes wealthier, this discrepancy will narrow, resulting in a "depreciation" of the PPP exchange rate.

We projected market exchange rates as follows: We first estimated the current level of "undervaluation" by comparing the market rate with the PPP rate. Currently, the ratio of the market rate to the PPP rate is 4.3. A more typical rate for a middle-income developing country is 2 or less. Over the course of the forecast period, the ration between the PPP and market rate should gravitate toward this ratio. Drawing on previous work by William Overholt, a coauthor of this study, we projected that China's real effective exchange rate (i.e., taking into account differences in inflation rates between the United States and China) will rise at an average annual rate of 2.3 percent per year over the next two decades. This implies that between 2003 and 2025, the renminbi will have appreciated by 66 percent in real effective terms against the dollar.

We then created a weighted average of these two projected rates by splitting China's economy into two components: goods and services that are traded and those that are not. This split corresponds to value-added from industry (traded goods) and the rest of the economy (non-traded goods and services). We then projected GDP by sector of origin for four sectors: industry (traded goods), agriculture and other primary economic activities, construction, and services (nontraded goods and services). The projections for the various sectors were constrained to generate increases in aggregate output equal to those in our projections of overall growth in GDP. We then converted each output series into dollars using the appropriate exchange rate: the projected market rate for value-added from industry and the projected PPP rate for value-added from the other economic sectors.

Under these assumptions, we project China's GDP in 2025 to run $9.45 trillion in 2001 dollars—a little less than half of projected U.S. GDP in 2025 but 94 percent of U.S. GDP in 2002 (Table 2.4). In 2025 per-capita GDP is projected to be $6,457 in 2001 dollars, 11 percent of the projected U.S. level and a little less than one-fifth of current U.S. average per-capita GDP.

By explicitly adjusting for the current "undervaluation" of the renminbi at the official exchange rate and for the effects that rising incomes—and hence wages—will have on the PPP exchange rate, this approach eliminates the biases introduced by assuming that the current

PPP and market exchange rates will remain static. For these reasons, we argue that this forecast of China's GDP using a combination of projected market and PPP exchange rates is the most useful for comparing future Chinese output with U.S. GDP. It is also most appropriate for comparing defense expenditures. We will employ this method to compare our forecasts of future Chinese military expenditures, developed in Chapter Seven, with expenditures on defense by the United States.

CHAPTER THREE
Government Revenues and Expenditures

Introduction

As discussed in Chapter Two, China has been one of the more recent and more successful of the Asian countries enjoying high rates of growth in the post–World War II period. Rapid economic growth has had salutary effects on the Chinese budget. Since the beginning of the reform in 1978, China's rapid rate of growth has dramatically lifted tax revenues. Tax revenues rose by 433 percent in real terms between 1978 and 2001, an annual rate of growth of 6.6 percent, reaching RMB 1,638 billion in 2001 or $198 billion (Figures 3.1 and 3.2).

Government expenditures have more than kept pace with rising revenues. They rose 505 percent over the same period, an annual rate of increase in real terms of 7.3 percent. In dollar terms they grew from $66.7 billion in 1978 to $228.4 billion in 2001, using the official exchange rates. Rising expectations and demands for government spending on infrastructure, education, health, and—more recently—social support for workers who have lost their jobs or people who have otherwise had their lives disrupted by the shift to a market economy have driven the rise in expenditures. The fact that expenditures have risen more rapidly than revenues has, not surprisingly, resulted in a widening budget deficit.

Pressures for increased expenditures are getting stronger. Demand for government services, especially on the part of the elderly, are rising. The Chinese government faces a host of expanding liabilities: A growing collection of unpaid bank debts, unfunded pensions, emergent social security obligations, and government guarantees constitute a formidable load. Moreover, not all the past increases in revenues have

51

Figure 3.1
Budget in Constant Prices

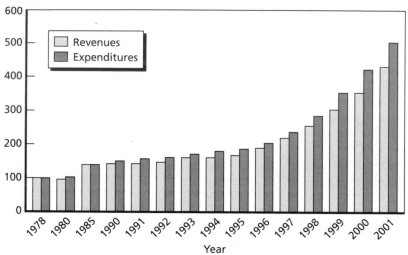

SOURCE: *China Statistical Yearbook,* various years.
NOTE: 1978 = 100.

RAND *MG260-3.1*

accrued to the central government: The revenue split between the central and local governments has waxed and waned, at times leaving the central government facing severe revenue shortfalls despite solid overall growth in total tax revenues in China.

To evaluate the current and future ability of the Chinese government to increase military expenditures, we first examine current sources of revenues and patterns of expenditures in China at both the national and local levels. Subsequently, we assess likely future government obligations and the revenues that the central and local governments are likely to be able to extract to deal with these obligations.

Information About China's Budget

Chinese statistical data are of varying quality and are often difficult to interpret. However, since the reform process got under way in 1978, statistical weaknesses frequently do not reflect deliberate deception on the part of the national Chinese government. Rather, they reflect the difficulty of collecting accurate statistics in a large, diverse, and

Figure 3.2
Total Revenue and Expenditures

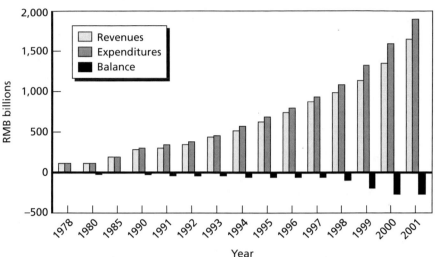

SOURCE: *China Statistical Yearbook*, various years.
RAND *MG260-3.2*

still poor economy. An additional complication pertaining to data on China's budgetary expenses and revenues is that the accounting system has frequently changed as it has evolved from a Soviet-style system to one more appropriate for a market economy. In many cases the accounting treatment has changed every three to four years. With each change, one would hope for a new, consistent time series, but that is frequently not the case. The tables and graphs in this chapter, mostly derived from the *China Statistical Yearbook,* must be understood in this light. Thus, the data are frequently not as comparable from one year to the next as the graphs may make them appear.

China's Budgetary Policy History

The Pre-Reform System

Since the Chinese government began to abandon central planning in 1978–1979, the government has struggled to adapt taxes and budgetary policies to an increasingly market-oriented system. The Chinese phrase for taking a job in the private economy is "jumping into the sea." Chinese budget officials during this period have in effect jumped into a stormy sea with a powerful rip tide.

Prior to reform, the Chinese government used two means to generate revenues for government expenditures: taxes and sequestration of state-owned enterprise (SOE) profits. During the pre-reform period, the accounts of the SOEs were treated as part of China's consolidated budget.[1] "Profitable" SOEs were the source of funds that the government then used for subsidizing loss-making SOEs, for investment, or for more traditional government expenditures. During this period, sequestration of enterprise revenues and "profits" frequently contributed more to the state budget than taxes did. For example, in 1978, enterprises provided just over half of total government revenues.[2] Enterprises were also the collection point for more traditional taxes, such as excise and industrial-commercial taxes, a combination of revenue (turnover), sales taxes, value-added taxes (VAT), and royalties. Enterprises collected the bulk of tax revenues. The Chinese government also levied excise taxes on goods such as salt, alcohol, and cigarettes, but these were also collected by the producer. Remaining taxes consisted of tariff revenues, agricultural taxes, and corporate income taxes levied on collectives and cooperatives, entities that fell outside the state-owned system.

Because the state controlled prices of both input and output from SOEs, decisions by the national government's price-setting agencies and government decisions on the allocation of resources in short supply were often the most important determinants of enterprise profitability. Reported profits in this system did not reflect real value-added or the efforts of the enterprise managers to improve operations, nor did they indicate worthwhile areas in which to invest. Not surprisingly, when price controls were phased out, many formerly "profitable" enterprises generated nothing but losses. This system of revenue collection was antithetical to encouraging managers to pursue profits. Profit sequestration and subsidization eliminated the profit motive as

---

[1] Prime, 1991, p. 169.

[2] *CSY,* 2002, Table 8.3.

an incentive for improving enterprise efficiency because managers were neither rewarded for profits nor punished for losses. The inefficiencies inherent in this system were a key cause of the collapse of the Soviet Union. Chinese leaders were concerned that they could lead to the collapse of their system as well.

Despite the centralized nature of this system, most revenues, both taxes and sequestered profits, were collected at the local level. In 1978, local governments collected 84.5 percent of China's total budgetary revenues, primarily through the enterprises located in their jurisdictions.[3] This breakdown did not reflect local control so much as the central government's inability to set up a national system of tax collection. Because of limited coverage and the poor state of China's communications and transportation networks, the central government found it more efficient to devolve revenue collection to the provinces. Expenditures, on the other hand, were more evenly divided, in part because the central government played a very large role in determining investments. For example, in 1978, the central government accounted for nearly half of all expenditures, 47.4 percent, even though local governments were responsible for providing the bulk of government services such as education and health care.[4] Because provinces collected the vast majority of revenue but the central government decided how most of these revenues were to be spent, there was a mismatch between provincial and central government incentives. The incentives for provincial authorities to increase tax revenues were limited because such a large share of these revenues was sent to Beijing.

In this centrally planned system, the People's Bank of China, the national bank, played a key, albeit passive, role. It served as the "ATM" for the Ministry of Finance, providing loans on demand. There was no need for credit analysis because lending decisions were made by the government planning apparatus. Loans were frequently treated as grants that were never intended to be paid back. The process was simple and centralized, and it appeared to work. We say "appeared" because the natural trajectory of such a system is toward larger and larger subsidies

---

[3] *CSY,* 2002, Table 8.13.

[4] *CSY,* 2002, Table 8.14.

as less efficient enterprises absorb increasing amounts of resources that would have been better invested elsewhere in the economy.

Reform and Taxation

As the economy became decentralized, this system began to falter. In 1980, the Chinese government embarked on the first of a number of tax reforms. In this first reform, the government divided responsibility for expenditures between local governments and the central government. It also began to take steps toward reducing the role of profit sequestration in government revenues. By 1983, four years after the beginning of the reform, the Chinese national government had virtually abandoned profit sequestration, replacing it with corporate income taxes, among other more market-oriented taxes.[5] Step by step, the government has pieced together a more workable system, but a system that retains so many ad hoc qualities that Western analysts marvel that it works at all.

In 1985, the government introduced a tripartite system dividing tax revenues into three separate pots: monies that accrued only to the central government, those that accrued to the local governments, and shared revenues.[6] In addition, provinces with surpluses were required to transfer funds to provinces with deficits—a system with blatant disincentives to sound provincial fiscal management. If a governor did a good job and ran a surplus, he had to give money away; if he spent the money wastefully, his province was likely to receive transfers from other provinces.

Effectively, China organized its finances as if it were a federal system rather than a unitary nation. This federal system, with all sorts of ad hoc measures to bind it together, has enabled Beijing to cope with the complexity of China's vast and diverse society, but has also introduced severe constraints and inconsistencies. Grasping this federal complexity and its costs is the key to understanding how China's budget process shapes Beijing's future strategic options.

---

[5] Prime, 1991, p. 170.

[6] This account of early changes relies heavily on Agarwala, 1991, especially pp. 9–10; and Prime, 1991, p. 170.

After the 1980 tax reforms, the Chinese government continued to experiment for a decade and a half, trying to devise a workable system. An ongoing problem has been the tension between the desire of the federal government to ensure that adequate revenues are raised to cover budgetary expenditures and China's desire to spread the wealth by transferring revenues from wealthier to poorer regions. In 1988, the government again changed the system of dividing tax revenues between the provinces and the central government. It created six different systems tailored to six different groups of provinces. Each system provided the provinces in that group with incentives to raise tax collections by allowing them to retain a large share of any increased revenues. Along with enabling both the central and provincial governments to fund their basic tasks, this ad hoc system achieved a redistribution of expenditures so that the poorer provinces spent about the same in per-capita terms as the richer provinces.[7] However, having six different revenue systems was a bureaucratic nightmare.

This fumbling reflected a series of problems. With economic reform, the sources of revenue were changing. As prices were liberalized and competition was introduced, state-owned enterprises were no longer guaranteed generators of profits and hence tax revenues. Increasingly, economic activity and potential tax revenues were being generated by cooperatives or township and village enterprises. However, the central government continued to depend heavily on the provinces for tax collection, and the provinces naturally preferred to keep the money at home. Conditions in every province were different, so ad hoc arrangements had to address the differing local situations. Because the provinces had been allowed to keep a disproportionate share of any increase in revenues as an incentive to collect more taxes, over time the central government's share declined relative to the provinces. And because the rules created all kinds of undesirable outcomes, both the central government and the provinces ignored the rules much of the time and introduced ad hoc fixes. That this worked at all was a tribute to the competence of the officials and the cohesion of the country, but such a Rube Goldberg system could not be sustained over the long run.

---

[7] Agarwala, 1991, p. 16.

As traditional sources of revenue declined, and as the center became more dependent upon reluctant provinces for the collection and submission of revenue, growth in the central government's revenues slowed (Figure 3.1), and its share in total government revenues collapsed both as a share of the total and as a share of GDP. As Figure 3.3, shows, total government revenues shrank from 32.2 percent of GDP in 1979 to 10.7 percent of GDP in 1995.

In 1994, this precipitous decline led to a new central-provincial budget deal, the Tax Sharing System or TSS. Under the TSS, the center would get 60 percent of total revenues rather than the roughly 40 percent it had been getting, but the center guaranteed each province that its revenues would never be less than what it had received in the year before the deal was put in place. The natural result was a great scramble by the provinces to collect enough revenue to establish a very high base. As a consequence, central government tax revenue as a share of GDP actually fell in 1995 and 1996 as the Chinese central government adhered to its commitment to the provinces. By 2002, however, the central government had succeeded in increasing its tax revenues to 18.5 percent of GDP.[8] Furthermore, the central government's share of total tax revenue increased.

The successful shift in tax revenue from the provinces to the central government in only a few years was due to the following factors:

- Better incentives for the provinces to collect taxes
- More decisive enforcement of tax laws by the central government
- Refinement of tax collection mechanisms
- Improved state enterprise profitability (especially reduced losses)
- Rising foreign investment
- Rising retail trade, and hence higher sales and value-added tax collections
- Removal of the military from the majority of the businesses that the PLA had owned, thereby making them liable for tax (PLA-owned businesses did not have to pay corporate income tax.).

---

[8] Both revenue and expenditures are a bit larger than the Chinese statistics show, because those statistics treat state enterprise losses as negative revenue rather than as state expenditures.

Figure 3.3
Central Government Revenues as Share of GDP

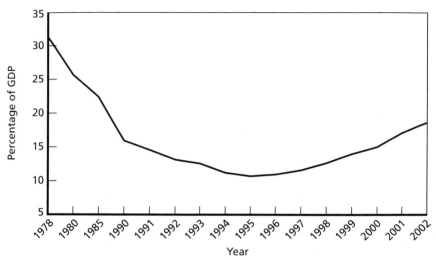

SOURCE: For 1978–2001, *CSY*, Table 8.2; for 2002, Finance Minister's Annual
Budget Report for 2002 (March 6, 2003).
RAND *MG260-3.3*

Rising foreign investment was particularly important, since foreign-owned companies largely pay their taxes while Chinese companies attempt to avoid them. In addition, Beijing has begun to limit the localities' rights to exempt from taxation companies that invest in their areas. Previously, every province and city competed to set up special economic development zones and attract foreign investors by exempting them from taxes. The provinces were delighted to give away Beijing's tax revenues in return for investments that provided local jobs.

The PLA was ordered to divest itself of its holdings of commercial enterprises at the 15th Party Congress in 1997. The divesture of PLA-controlled companies has had an important impact on budgetary revenues. Through the use of military ports, airfields, and trucking fleets, the PLA and PLA-controlled companies were able to smuggle very large volumes of goods across China's borders and distribute them throughout the country. Both domestic and foreign investors were able to use the umbrella of the PLA to avoid paying taxes other than tariffs—and often they avoided paying tariffs, too. Customs revenues jumped 80 percent between 1998, the year the PLA began divestiture, and 1999. Divestiture definitely contributed to this increase.

Trends in international trade and foreign investment are dispro-portionately important for tax revenues, because it is much easier to collect taxes from foreigners than to collect them from Chinese. Just how important is evident from a report by the tax authorities in March 2003 that revenues were up 28 percent over the corresponding period in 2002, in part because of a 66.5 percent increase in import duties col-lected.[9] What is happening here is a virtuous circle: As tariff rates de-cline, traders are more willing to pay them, collection efforts increase, and smuggling decreases. The reality of Chinese tariffs may be the op-posite of what it appears: The average effective rate may be rising rather than falling, because in the past the army smuggled such vast quantities of goods that much of the time the effective tariff rate was zero.[10]

The 1994 tax reforms had a substantial impact on tax revenues. Tax revenue as a share of GDP rose—albeit to a little over half the share of GDP it had been prior to the beginning of reform. In addition, the central government's share of total tax revenues rose dramatically (Figure 3.4). The government's success in raising revenues has come in the face of some trends that created fears of relative decline. In his 2002 budget speech, Finance Minister Xiang Huaicheng argued that the rapid rise in revenues as a share of GDP was coming to an end and forecast growth in revenues in 2002 to be in line with GDP. He noted a number of causes of a potential slowdown or even a decline in tax revenues:

- The average tariff rate fell from 15.3 percent in 2001 to 12 per-cent in 2002 following China's accession to the WTO.

---

[9] "Tax revenue up 28 percent by March 20," China News Service, Quoting Xie Xuren, Direc-tor of the State Administration of Taxation, carried in ChinaOnline.com, April 1, 2003.

[10] The extent of smuggling through the PLA is best illustrated by sales of liquor distilled in North America. Although trade statistics show negligible or no imports of liquor from North America, products of these companies are available on virtually every major street in every ma-jor city in China. These products had been smuggled into China by the PLA. The PLA's vast distribution system ensured availability throughout the country. This widespread smuggling deprived the Chinese government of revenue and has seriously distorted trade statistics.

Figure 3.4
Central and Local Shares of Revenue

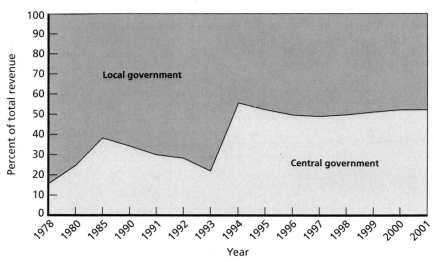

SOURCE: *CSY*, Table 8.13.
RAND *MG260-3.4*

- Stamp tax rates on stock transactions were reduced from 4 percent on A-shares and 3 percent on B-shares to a unified rate of 2 percent.
- Business tax rates on banking and insurance fell from 7 percent to 6 percent.
- The government ceased sales of shares in state-owned enterprises on stock markets.

Minister Xiang was too pessimistic. His March 2003 budget report reviewed the outcome: "For a time, the situation was pregnant with grim possibilities for budgetary revenues and expenditures, especially those in the central budget."[11] However, revenues continued to rise as a share of GDP in 2002 and 2003.

Although the increase in the share of total tax revenue raised by the central government suggests that the 1994 tax reform achieved one of its goals, the central government does not, in fact, spend much of

---

[11] Xiang, 2003.

the revenues it takes in but transfers them to the provinces because most expenditures are made by local governments, not by Beijing.

The World Bank has put together comparative data on central government revenues from different countries using common definitions. The results are shown in Figure 3.5. Even by comparison with the United States and the Russian Federation, two other large countries with decentralized systems of government, China's finances are very decentralized. The Chinese numbers have implications. If the Chinese government wished to alter budget allocations quickly because of a crisis or wanted to have greater control over revenues, French President Jacques Chirac's position would be preferable to that of China's Hu Jintao.

Figure 3.5
Central Government Revenue and Expenditures

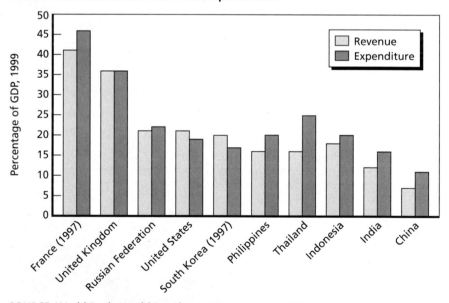

SOURCE: World Bank, World Development Indicators, 2002, CD-ROM version.
NOTE: Data for France and South Korea not available for more recent years.
RAND *MG260-3.5*

Tax Revenues

Division Between Central and Local Governments

The central government's revenues come largely from large state-owned enterprises, customs duties, excise taxes, and value-added tax (VAT) (Table 3.1). VAT has become increasingly important to the central government, in part because it receives three-quarters of VAT receipts. Local governments receive tax revenues from corporate profit taxes, personal income taxes, property taxes, agricultural taxes, and natural resource taxes plus their 25 percent share of VAT. Since taxes on trade, foreigners, and state enterprises are easier to collect, for the time being the central government finds it easier to collect taxes than the provinces or municipalities. But in the longer term, revenues from personal and corporate profit taxes are likely to grow more rapidly than revenues from tariffs and taxes on state-owned enterprises. In other words, in the longer-run the division of revenue sources appears more favorable to local government than the central government unless the system is changed again.

Table 3.1
Sources of Government Revenues

Central Government Revenues	Local Government Revenues
Customs duties, excises, and VAT levied on imports	Business taxes except those collected from banks, railroads, and insurance companies
Excise taxes	25 percent of VAT and 9 percent of the securities stamp tax
Income tax collected from central government–owned enterprises	Corporate profit taxes apart from those collected from central government–owned enterprises
Income tax collected from railroads, headquarters of banks, and insurance companies	Personal income taxes
75 percent of VAT	Resource taxes
91 percent of the securities stamp tax	Urban maintenance and construction taxes, real estate taxes, agricultural taxes, and a number of small taxes

SOURCE: Zhihua Zhang, Minister of Finance, Intergovernmental Fiscal Relations in P.R. China, presentation for World Bank.

Detailed Sources of Central Government Revenues

The largest source of central government revenue, over one-third of the total, is VAT (Figure 3.6).[12] Combined with the consumption tax, the two account for 47 percent of total central government revenues. The other major source of central government revenue is from foreign trade: Tariffs (10 percent of total revenue) plus VAT and consumption taxes on imports (20 percent) provide more than a quarter (30 percent) of total revenues. The foreign sector is even more important than these figures suggest. A disproportionate share of corporate profit taxes and personal income taxes is paid by foreigners. The Chinese government has found it easier to collect taxes from foreigners than from domestic sources and therefore relies heavily on the more vulnerable and compliant foreigners.

Figure 3.6
Central Government Revenue Sources, 2001

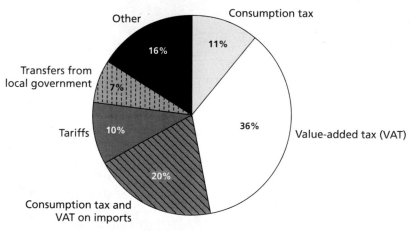

SOURCE: *Finance Yearbook of China 2002*, p. 296.

RAND *MG260-3.6*

---

[12] Figure 3.6 reflects important modifications to the Chinese statistics. The government statistics treat subsidies to state and collective enterprises as negative revenues, but they are more properly regarded as expenditures. Thus we have removed them from the revenue accounting. In addition, the yearbook treats rebates to exporters as negative revenues; we have instead netted them out of value-added taxes. The graph uses revenues actually collected, not planned revenues.

Personal income taxes provide a relatively small contribution to central government revenues (3 percent of the total). The income tax threshold is relatively high so most Chinese families are not liable for income tax. Most that are (and many businesses) simply do not pay. In the mid-1990s, only a handful of employees in People's Bank of China, the central bank, bothered to pay their income taxes.[13] There is enormous room for improvement here.

Local Government Revenues

Figure 3.7 decomposes local government funds by source.[14] Nearly half (42.7 percent) of those revenues consist of transfers from the central government. The central government is highly dependent on local governments to collect tax revenues, however, even though it now has the dominant claim on VAT and so becomes the listed source. This some-

Figure 3.7
Local Government Revenue Sources, 2001

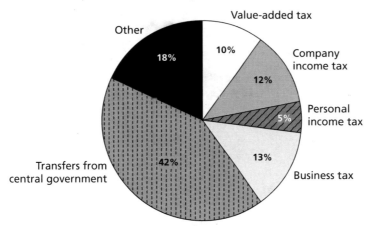

SOURCE: *Finance Yearbook of China 2002*, p. 296.
RAND MG260-3.7

---

[13] Information provided in conversations with People's Bank of China staff.

[14] Figure 3.7 reflects the same modifications as Figure 3.6.

what circular situation should not be allowed to disguise the heavy dependence of the central government on the local collection network.

Business taxes (13.1 percent) and corporate income taxes (12 percent) together provide a quarter of total local revenues. The importance of this revenue source will presumably grow as private business expands and collection efforts improve. However, local businesses are also burdened by fees. The burden is so heavy that increases in revenues from corporate taxes will likely come at the expense of fees, with little net gain in total government revenue, but a shift from off-budget to budget revenues. Personal income taxes provide 5.1 percent of local revenues. Net gains from these taxes will probably have to come from wealthy, largely urban, individuals rather than rural dwellers who are the most heavily burdened by local fees and taxes, especially for education and health services.

The national figures hide a wide range of variation among provinces. Figure 3.8 shows the distribution of revenues for one of the richest provinces, Jiangsu, in 2001. Figure 3.9 shows the same graph for one of the poorest, Gansu. In 2001, per-capita income in Jiangsu was RMB 12,922 while in Gansu, it was RMB 4,163, about one-third of Jiangsu's.

Figure 3.8
Jiangsu Revenues

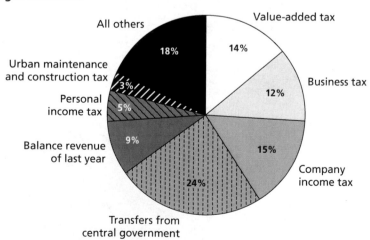

SOURCE: *CSY*, various years.

RAND *MG260-3.8*

Figure 3.9
Gansu Revenues

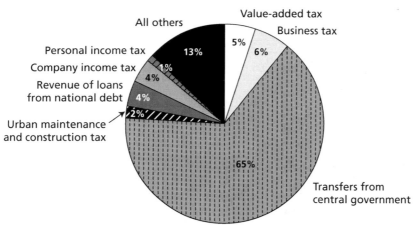

SOURCE: *CSY*, various years.
RAND *MG260-3.9*

As China continues to develop, the distribution of tax revenues in the poorer provinces is likely to gravitate to the pattern of Jiangsu. This wealthy province is only about half as dependent on transfers from the central government as the national average. Its corporate income tax is in line with the national average.

Almost two-thirds of local budget revenues consist of transfers from the central government in Gansu, compared with one-quarter in Jiangsu. The only other substantial items are VAT, the business tax, and corporate income tax revenues, 5 percent, 6 percent and 4 percent of the total, respectively, along with "Other Revenue." Revenue of loans from national debt finances a fairly substantial share of expenditures as well. Personal income tax provides only 1.4 percent of total revenue. The tiny increments provided by items like personal income tax might suggest room for improved collection, but for the most part people are so poor they fall below the threshold for paying income tax. In short, Gansu is dependent on central handouts.

This province, along with another impoverished area, Tibet, is one of the two provinces where China's new leader, Hu Jintao, had governing responsibilities. His belief that the government must drastically increase the priority of helping such areas out of poverty, compared with a single-minded focus on rapidly increasing total GDP, reflects a major policy difference between his administration and the previous one.

The Large and Vital Role of Extra-Budgetary Revenues

As noted above, total government revenues of 17 percent of GDP are quite small by both developed-country and even medium-income developing-country standards. This figure gives the impression that there is a great deal of room for raising the share of budgetary revenues in GDP as the Chinese economy. But a great deal of what the Chinese people pay for government services is off-budget. Figure 3.10 shows official government estimates of on- and off-budget revenues.

The government figures indicate that extra-budgetary revenues are 28.6 percent of on-budget revenues and 22.2 percent of total revenues, on- and off-budget—i.e., a little over one-fifth of total government revenue. According to World Bank surveys, off-budget revenues

Figure 3.10
On- and Off-Budget Revenue: Official Figures

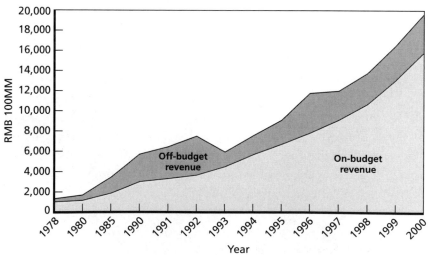

SOURCE: *CSY*, Tables 8.1 and 8.18.
RAND *MG260-3.10*

are actually far larger.[15] Chinese citizens pay a substantial amount of income in fees and other charges to local governments. It is not surprising that government statistics miss some of these revenues, because so many of the fees and charges are ad hoc and in some cases technically illegal. Some of the largest payments by poorer households are fees and charges for education and medical care. Both services are mandated by the central government, but local governments are not given adequate revenues to fund these services so local clinics and schools have to augment their up revenues, usually by levying fees. These figures also do not take into account bribes to officials: virtually every transaction requires a bribe in China.[16] Bribes are usually substantial in relation to the transactions themselves. In short, if one could somehow get adequate data covering household payments for education and health and bribes, total payments to the government or to government employees would exceed reported government revenues, probably by a significant margin. In other words, households are paying a higher share of their incomes to the government or government officials than official statistics suggest, implying that the ability to raise tax revenues by increasing taxes or expanding the tax base is less than it appears, unless these taxes are used to replace current payments for government services such as fees for education and health care.

The localities prefer to collect payments for education and health care in the form of fees rather than to raise tax revenue, because Beijing takes a share of on-budget revenues and imposes constraints as to how the remainder of on-budget funds can be spent. Off-budget funds, on the other hand, stay at home and their use is unrestricted. Beijing is trying to restrict the extra-budgetary revenues, because arbitrary fees deter business investment in the cities and are frequently perceived as abusive or oppressive, especially by farmers.

Some extra-budgetary funds come from "dividends" or transfers paid to local municipalities by municipally-owned enterprises. Early in

---

[15] World Bank 2002, p. 14ff.

[16] This assertion is based on innumerable conversations with foreign and domestic businesspeople operating in China. The writer of this chapter lived in Hong Kong working for investment banks for 16 years and served on the boards of the most important association of foreign businesses and one of the most important local associations.

China's economic reform the central government reduced transfers to local governments from central government funds, but in return permitted them to profit from enterprises that the municipalities created. The result was a wave of entrepreneurship, particularly as rural officials founded town and village enterprises. By 1996, these enterprises employed 135 million people, primarily in rural areas. But as the reform era has matured, the TVEs have encountered more competition from reformed state enterprises, private and foreign-owned enterprises, and imports. Consequently, employment in TVEs has fallen.

TVEs quickly became more important as employers of local residents and sources of extra income for village and party officials than sources of additional revenues for local governments. From their initiation, most of these enterprises were operated as if they were privately owned: village officials became the new capitalists. At first, some of these enterprises were very profitable and provided local authorities with significant additional resources for local government expenditures. However, because local officials retained enterprise earnings for their own use, the extent to which TVEs provided substantial off-budget revenues for expenditures on local government services has been limited. Now that many of these enterprises are losing money or are barely profitable, their ability to serve as a source of local government revenues is even more circumscribed. As a result, they have not been a reliable, expanding source of budgetary support.

Locally collected extra-budgetary funds create problems for the central government. They are beyond the central government's control and they diminish Beijing's ability to extract additional resources from the population. They are often quite arbitrary: as soon as a business starts making money, the locality invents a range of new fees to extract some of the profits. Such arbitrary extractions both damage the prospects of local entrepreneurs and alienate foreign investors. The numbers and scope of fees is often stunning. For instance, the *People's Daily* reported that on April 9, 2003, in Beijing, "The vice-mayor also pledged . . . that the municipal government will scrap 175 categories of administrative charges for foreign-funded companies by the end of this year."[17] In most other countries it is unimaginable that there would

---

[17] "Reform of Administration Piloted in Chinese Capital," *Peoples Daily,* April 10, 2003.

be 175 categories, never mind the possibility that one could scrap 175 and still have any left over.

In rural areas, the fees are extremely numerous and burdensome, sometimes leading to demonstrations, riots, and even murders of local officials. Anecdotal accounts suggest that the fees levied on villagers reached the level of a social crisis in 1999–2000. Premier Zhu Rongji launched an experiment in Anhui designed as a pilot program for re-placing numerous fees with a normal tax system; the project failed miserably, and Premier Zhu, successful in so many other areas, sub-sequently became a target of bitter criticism. At the National Party Congress in March 2002, he declared that nothing gave him such headaches as the problems of the farmers. In 2003, the new Premier, Wen Jiabao, broadcast his determination to implement a nationwide program of swapping fees for taxes to reduce the burden on farmers, but many fear that the effort will force local governments to curtail ex-penditures on education and other important services.[18] In some poor and rural areas, financial inability to hire and retain teachers has cre-ated local educational crises, and in considerable swaths of rural China it has created a squeeze.

Another promised reform is separation of the revenue collection function from the expenditure function. That separation is intended to reduce the ability of officials to invent new fees whenever they need to fund a new project. Some scholars think it will reduce the government's ability to collect taxes by reducing local governments' incentives to do so.[19]

Government Expenditures

The Central-Local Division of Responsibilities
Although Beijing's share of revenues has risen, its share of expenditures has stayed low and is tending to gradually decline. In 2002, the locali-ties accounted for 71.5 percent of total government expenditures, not

---

[18] See Josephine Ma, "Farmers a Poor Second in Wage Stakes," *South China Morning Post,* April 10, 2003.

[19] Lin, 2003, p. 84.

counting payments on the national debt, while the central government accounted for only 28.5 percent, down from 30.3 percent in 1997, according to the Finance Minister's 2003 Budget Report (Figure 3.11). The share of total expenditures made by the localities is high and rising.

Of the 71.5 percent of expenditures made below the central level, most of the expenditures are made below the provincial level: Counties and townships account for 55 percent of all government expenditures in China.[20] The number of entities involved is huge: Local governments include 31 provinces and province-level cities, 331 major cities, 2,109 counties, and 44,741 townships.

Local governments account for the lion's share of expenditures because they are responsible for almost all expenditures on health, education, social services, social security (including unemployment insurance), local infrastructure, and agricultural development. (See Figure 3.12.)

Figure 3.11
Central and Local Shares of Expenditure

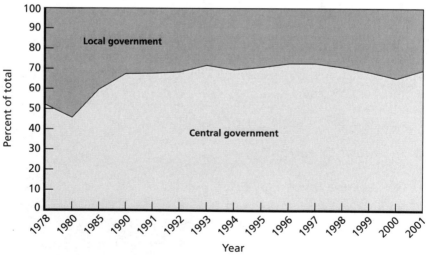

SOURCE: *CSY*, Table 8.14.
RAND *MG260-3.11*

---

[20] Wong, 2001, p. 1.

Figure 3.12
Expenditures, Central and Local

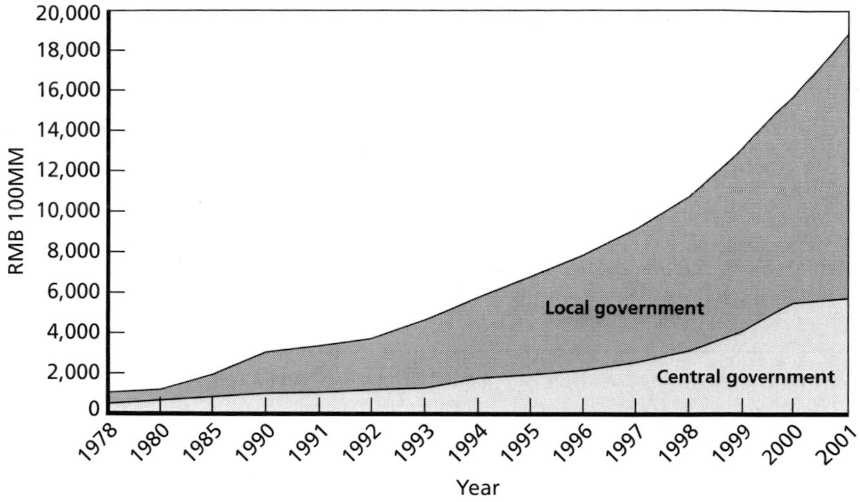

SOURCE: *CSY*, Table 8.14.
RAND *MG260-3.12*

Although there are considerable overlaps in responsibilities, the local governments bear primary responsibility in the areas of overlap. For instance, the central government is responsible for the construction of the few national roads and for the railroads, but local governments are responsible for building most roads, schools, and clinics. Local governments, in fact subprovincial governments, bear most of the responsibility for education, health, pensions, and unemployment compensation (see Table 3.2).

Local governments have been forced to bear the costs of supporting or restructuring state-owned enterprises in their area. When state enterprise losses burgeoned, Beijing announced that all but the biggest enterprises would henceforth be the responsibility of local governments. Because of the pending losses, local governments promptly sold off as many as they could, notwithstanding the official national policy against privatization. However, local governments still have significant residual responsibilities.

The Chinese central government assigned the localities tasks that are often the responsibility of the central government in other countries.

Table 3.2
Government Responsibilities for Expenditures

Central Government Responsibilities	Local Government Responsibilities
Foreign relations	Local governmental administration
National defense	Domestic security (dividing expenditures
Central administration	with the central government)
National capital construction	Updating technology of local government-
Updating technology for central	owned enterprises
government-owned enterprises	Local agriculture development, subsidies
Debt payments	Urban maintenance and construction
Limited responsibilities for agriculture,	Local culture, education, health, and most
education, health, science, etc.	social safety net expenditures

SOURCE: Zhihua Zhang, Ministry of Finance, Intergovernmental Fiscal Relations in P. R. China, presentation for World Bank, undated.

The cities are responsible for most social security, unemployment insurance, and social welfare payments, while counties and townships pay for 70 percent of education and 55 to 60 percent of health expenditures.[21] All levels have heavy responsibilities for the construction of infrastructure. The localities often face a mismatch between fiscal responsibilities and revenues. They face official restrictions on levying taxes, but are assigned a wide range of responsibilities by the central government. Transfers from Beijing are often inadequate to cover these responsibilities.

Local governments have limited means to cover budgetary shortfalls. They are not allowed to borrow. Theoretically, they must finance the construction of major infrastructure projects from current revenue streams, although, as noted below, in practice they find ways to defer liabilities. At China's current level of development, there is a good reason for this prohibition. All lower levels of government have demonstrated a tendency to fiscal imprudence—from the irresponsible borrowing of the provinces' international trust and investment companies to the spectacular collapse of township financial institutions.[22]

---

[21] From Wong, 2002.

[22] In 1998, the Guangdong International Trust and Investment Company, the largest and most prominent of the provincial investment companies, went bankrupt with billions of dollars of debt after using borrowed foreign money for widespread real estate speculation, other imprudent investments, and outright corruption. It would be difficult to find an extensive network of financial institutions anywhere in the world that has been as badly mismanaged as China's credit cooperatives, which are managed at the township level.

Other peculiarities of China's fiscal system lead to perverse local outcomes. The central government effectively insures personnel costs but not other costs. If the local government is short of money for wages, the central government steps in and covers the payroll. Consequently, local governments have become bloated with people while being short on funds for nonpersonnel operating expenditures. These expenditure strictures have contributed to the growth in off-budget fees, which are used to cover nonpersonnel operating costs. In some instances, local governments lapse into periodic insolvency as they delay payments to suppliers, health care providers, pensioners, and teachers or fail to pay altogether. Shuanglin Lin cites studies showing that by 2000 some local governments in Hunan had spent their entire revenues through 2003 and some in Sichuan had already spent all expected revenues through 2015.[23] Overall, the budget structure is a peculiar combination of decentralized and centralized systems that in the end satisfies the requirements of neither.

Local Expenditures

Although China's 928,000 villages are not official levels of government, they still have important effects on budgetary expenditures. Each village employs both government and party officials. When things go wrong at the village level, riots ensue. The Communist Party has its historic roots in the villages, and for the majority of China's people the primary point of contact with the government and the Party is at the village level. Hence, the central government's legitimacy and stability can be threatened by adverse developments in the villages. In fact, corruption and incompetence at the village level became such a problem for the regime that it decided to allow relatively free village elections in an effort to put in place officials who would be competent and inspire local trust.[24] It has also considered abolishing the entire network of village governments, but so far has taken no concrete steps to do so.

---

[23] Lin, 2003, p. 89.

[24] According to independent U.S. observers, about three-quarters of village elections are actually conducted in a relatively free way. Multiple candidates contend, and the Communist Party's favored candidates frequently lose. This creates a potential problem for proposals to abolish village governments in the name of efficiency, because they may be perceived as having greater legitimacy than the township level above them.

Despite such measures as elections and required transparency of accounts, village administration and finances remain in poor shape. Official regulations forbid local governments from borrowing, but an official report estimated that in 2001 the villages alone had more than RMB 200 billion ($24.1 billion) of outstanding debts.[25] International financial institutions estimate county-level debt at 3–6 percent of GDP. Debts owed by local villages contributed to the collapse of local credit cooperatives at the end of the 1990s. They had lent heavily to the townships and villages.

Throughout China, late or unpaid salaries are a frequent occurrence. Many teachers and officials simply do not get paid for extended periods of time or are only paid in part. Finance Minister Xiang's 2003 Budget Report stated that "As of the end of 2002, 25 of the 36 provinces, autonomous regions and municipalities directly under the central government and cities specially designated in the state plan were paying wages and salaries on time and in full" and that in the remaining 11 arrears had dropped to RMB 1.7 billion ($205 million) from RMB 6.5 billion ($785 million) in 2001.

Overall, the finances of local governments have been deteriorating (Figure 3.13). The central government has required local governments to build most local infrastructure, and there has been a frenzy of road building and other infrastructure construction. Likewise, local governments have principal responsibility for education, and there is a national policy that every year education expenditures should rise faster than the total national budget.

More important, local governments have principal responsibility for a wide range of welfare responsibilities, and these responsibilities are rapidly becoming more burdensome. Under the old socialist system, pensions, unemployment payments, and medical services were the responsibility of the state enterprises and the communes. Now that those institutions have disappeared (the communes) or have suffered drastic declines in employment or revenues (the state enterprises), the services they formerly provided have devolved to local governments or in some cases are no longer provided. In many instances, these responsibilities were simply neglected by the local governments. In the early phase of re-

---

[25] *China Newsweekly* report quoted in Willy Wo-Lap Lam, "Chinese Villages Burdened with Massive Debt," http://edition.cnn.com/2001/WORLD/asiapcf/east/06/19/china.villages/.

Figure 3.13
Local Government Fiscal Balances

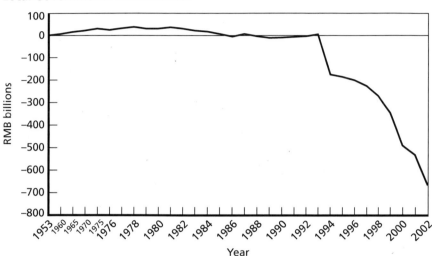

SOURCE: Lin, 2003, p. 83.
RAND *MG260-3.13*

form, that worked, because rural incomes were rising at very rapid rates. But lately that situation has been reversed: Rural incomes have stagnated and the demand for government services has increased rapidly.

As the state enterprises have gone into decline, unemployment has increased and the number of retired people whose pensions are not covered by state enterprises has risen sharply. China's health problems are also severe; many are far more urgent than the unemployment and pension problems. The SARS epidemic of the first half of 2003 dramatically exposed the dangers of allowing the rural medical system to collapse. And collapse it has, since the medical service provided by the state enterprises has not been replaced. Only about one-fifth of the population of high-income areas, and only 1–3 percent of the population of poorer areas, has access to cooperative medical facilities. The result is that China's economic miracle has recently coincided with a downturn in the people's health.[26] The failure to rebuild a rural health

---

[26] On the downturn, see World Bank, 1996. For medical coverage, see L. Zhu, "Who Can Provide the Farmers with Medical Services?" *Liaowang* (*Outlook Weekly*) (16 April 2000), cited in an unpublished draft paper by Tony Saich.

system became an acute social and political crisis with the onset of SARS; once it became clear that allowing rural SARS victims to travel to the cities for treatment was dangerous to society, there was often no choice other than to herd them into quarantine and allow them to die. Similarly, HIV is spreading rapidly in China and the health authorities have not demonstrated the ability to cope with such an epidemic.

Figures 3.14 and 3.15 show how the wealthy province Jiangsu and the impoverished province Gansu, respectively, spend their money. Education is the big item (15 percent), followed by government administration, urban maintenance, capital construction, and enterprise innovation. Social security, pensions, and the like are very small; health comes in at only 4 percent (included under "Other"). Keeping in mind that the provincial level has primary responsibility for social expenditures, this pattern shows investment in gross economic growth with little regard for the social consequences. It is fortunate that education is seen as a key to growth, but even as the biggest item in the budget, education's 15 percent is by no means large.

In Gansu, capital construction is the biggest item (15 percent), reflecting an emphasis on building roads and other infrastructure to

Figure 3.14
Jiangsu Expenditures

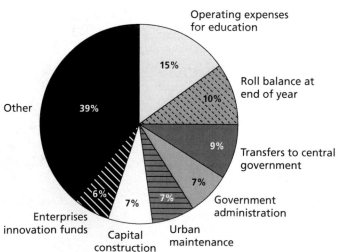

SOURCE: *CSY*, various years.

Figure 3.15
Gansu Expenditures

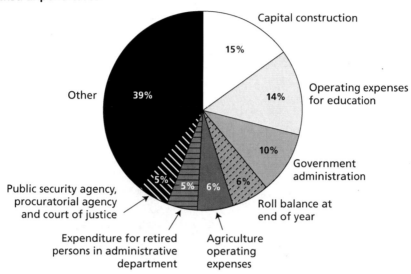

Capital construction

15%

Other 39%

Operating expenses
for education
14%

10%

Government
administration

Public security agency,
procuratorial agency
and court of justice

5%   5%   6%   6%

Roll balance at
end of year

Expenditure for retired
persons in administrative
department

Agriculture
operating
expenses

SOURCE: *CSY*, various years.

kick-start development and tie the province to the rest of China. Education follows just behind, with 14 percent, similar to Jiangsu but far less on a per-capita basis because the province is so poor. Government administration takes 10 percent, half again more than in Jiangsu because Gansu is both more backward and more socialist. Two categories of agriculture spending absorb 6 percent of the budget, reflecting a more backward economy than Jiangsu, where agriculture takes only 3 percent of total provincial expenditures. Pensions and social security get 4 percent each, welfare substantially less than 2 percent (also included under "Other"). In part, these small fractions of a small budget reflect the realities of an impoverished region that cannot afford too much welfare.

National-Level Expenditures

Turning to the national budget, Table 3.3 shows how different kinds of expenditure have evolved in both absolute and percentage terms. Some items have accounted for roughly the same share of total expenditures over the past 25 years; others have shifted markedly. For example, the

Table 3.3
Central Government Budgetary Expenditures by Category

Category	1978	1990	2001	1978	1990	2001
	RMB Billions			Percentage of Total		
Investment	45.2	54.7	251.1	42.8	20.6	19.7
Enterprise Operating Capital Investments	6.7	1.1	2.3	6.3	0.4	0.2
Science and Technology	6.3	15.4	99.2	6.0	5.8	7.8
Agriculture	7.7	22.2	91.8	7.3	8.3	7.2
Culture, Education, Health, and Science	11.3	61.7	336.1	10.7	23.2	26.4
Pensions and Social Relief	1.9	5.5	26.7	1.8	2.1	2.1
Government Administration	16.8	29.0	144.2	15.9	10.9	11.3
Price Subsidies	4.9	30.3	219.8	4.6	11.4	17.2
Defense	1.1	38.1	74.2	1.1	14.3	5.8
Other	3.8	8.3	29.9	3.6	3.1	2.3
Total	105.7	266.4	1275.3	100.0	100.0	100.0

NOTE: Percentages may not sum to 100 because of rounding.

government no longer provides enterprises with operating capital to any significant degree: Expenditures on state enterprise operating capital have plummeted from 6.3 percent of the budget in 1978 to 0.2 percent in 2001. This decline, however, is to some extent a statistical illusion because this burden has largely been shifted from the government budget to the state-owned banks. If the loans made to finance enterprise operations go bad (and many have), the government will have to pick up the tab. Investment has declined from over 40 percent of the budget to under 20 percent as enterprises and state-owned banks, not the budget, now finance corporate investment. Government investment, now almost exclusively in infrastructure, has fallen from 12.5 to 2.6 percent of GDP between 1978 and 2001. Central government expenditures on infrastructure are likely to remain in this neighborhood (2.5 to 3 percent of GDP) because China has such a long way to go in terms of building a modern, national infrastructure. The national government is expending considerable resources on research and development (science and technology promotion). The Finance Minister's 2003 Budget Report states that in real terms expenditures for science and technology rose 950 percent between 1997 and 2002. The share of the national budget spent on science and technology promotion rose from 6.0 to 7.8 percent between 1978 and 2001. Expenditures in this category are likely to continue to grow as a share of total expenditures in the future.

Expenditures on government administration have soared in the past 23 years along with overall growth in budgetary expenditures, quintupling between 1990 and 2001 despite a vast effort toward the end of that decade to cut the bureaucracy sharply. Indeed, drastic cuts in numbers of personnel, roughly halving government employment in the central government and in key provincial governments, have been almost exactly offset by doubling civil service salaries over the same period.[27] Salaries for civil servants had seriously lagged behind those in the private sector, with the prime minister making only a few thousand dollars annually just a few years ago.[28] Low salaries contributed to incompetence and corruption. Cutting the bureaucracy at the lower levels is much more difficult, because lower-level bureaucrats lack the skills and networks that assure their seniors quick employment in the private sector. The next few years are likely to see much more vigorous efforts to trim the bureaucracy at lower levels but with any resulting budget gains offset by rising salaries and severance payments to officials who have been laid off.

While expenditure reductions from cutting personnel are likely to be limited because remaining employees are receiving higher salaries, the potential for efficiency gains are certainly there. In 2002, the government began separating the administration of revenue collection from the administration of expenditure—thereby presumably reducing the arbitrary creation of fees and taxes whenever a new project required

---

[27] The Finance Minister's 2003 Budget Report states, "Per capita basic monthly salary of employees in government departments and institutions at the end of 2002 doubled that of 1998." Those numbers are in nominal figures; since this was a period of deflation, they more than doubled in real terms.

[28] According to the 1999 *Asiaweek* salaries survey, in 1998, "the base salary of China's president was US $103 per month; that of the governor of the central bank, $107; the Finance Minister, $82; and a typical clerk, $22 plus bonuses that could be several times that figures." See http://www.asiaweek.com/asiaweek/99/0305/cs2.html. It is obviously imperative for such compensation to rise fairly rapidly.

support. In addition, according to Finance Minister Xiang's 2003 Budget Report, the consolidation of expenditures into a centralized treasury function "reduced the number of links in the transfer of funds," saving 10 percent on total state procurement expenditures. Reduction in the number of offices and licenses, along with a reduction in corruption, could presumably improve government efficiency and release some funds for other purposes in coming years.

Hidden Budgetary Problems

The official story of the Jiang Zeming and Zhu Rongji years is that recent rapid increases in government expenditures have been possible with a budget deficit that has widened to only 3 percent of GDP, government debt that is only 16–18 percent of GDP, and foreign debt that is less than China's huge foreign exchange reserves. The last part is true. The other parts are true only in the narrow sense of an accounting fluke.

At 3 percent of GDP, the official budget deficit is at a level that can be sustained for a considerable number of years by a country whose finances are otherwise in good shape. However, the Chinese government has a very large stock of off-balance-sheet liabilities. Servicing these liabilities will become increasingly expensive in the coming years, as they revert to the Ministry of Finance and have to be acknowledged as official government debt. The IMF estimates that, if interest payments on these off-balance-sheet liabilities are included, the government's current budget deficit would run between 5 and 6 percent of GDP, not 3 percent of GDP.[29] Ultimately, even if interest rates continue to be suppressed and large liabilities carried with no imputed interest, the opportunity costs of these liabilities will become a drag on economic growth.

Liabilities Not Included in Central Government Debt
Nonperforming Loans. Beijing has limited the budgetary cost of keeping the big state-owned enterprises (SOEs) afloat by instructing the banks to lend to them to cover operating losses. As SOEs continue

---

[29] IMF, 2002, Chapter One, Box 1.4.

to lose money, the result has been a huge buildup of nonperforming loans (NPLs). The Chinese government estimates the stock of nonperforming loans at 27 percent of GDP. The IMF estimates total nonperforming loans at 50 to 75 percent of GDP, "including an amount equal to 15–1/2 percent of GDP that has been transferred to asset management companies."[30] Private analysts' estimates of bank NPLs cover a considerable range but fall within the bounds of the IMF estimates. All these numbers, even those of the Chinese government, are very large: An acceptable level of nonperforming loans in a developed market economy would be on the order of 2 percent of GDP.

The nonperforming loans are part of a gigantic financial shell game. The government pushes the SOE subsidies off-budget by delegating them to the banks, where they become nonperforming loans. Then it tries to save the banks by having them sell those NPLs to asset management companies at face value.[31] The asset management companies pay for the loans by floating bonds guaranteed by the central bank. Since the loans are worth only ten to twenty cents on the dollar, that effectively puts 80–90 percent of the transferred bad loans onto the balance sheet of the central bank. That in turn ultimately endangers the central bank's solvency.[32] Eventually the shell game must end by replacing the central bank's bad loans with government debt.

The Chinese government has yet to fully recognize the extent of this problem. Although it is selling off some of these bad debts, generally at a tiny fraction of face value, it is inhibited from accelerating the sales for fear of increasing layoffs. Sales are also inhibited and sale prices depressed by ambiguities regarding the property rights conveyed to those who purchase the assets.

Government Debt. Government domestic debt starts with overt debt equivalent to 16 percent of GDP. To this one must add loans from the World Bank and Asian Development Bank plus bonds issued to

---

[30] IMF, 2002, Chapter One, Box 1.4.

[31] Asset management companies are firms, like the U.S. Resolution Trust Corporation, that take over banks' bad loans and attempt to restructure the companies or sell off the loans so that the banks can move forward with a relatively clean slate.

[32] The most comprehensive analysis of this situation can be found in Ma and Fung, 2002.

recapitalize the banks. This adds up to 25 percent of GDP at market exchange rates.[33] To that one must add local government debts, not including provincial-level debt because it is in the form of nonperforming loans; a reasonable ballpark estimate would be 8–10 percent of GDP. Finally, one must add that the government will ultimately be responsible for most of the nonperforming loans of banks and other financial institutions, as well as for a substantial proportion of unfunded pensions.

Interest payments on central government debt are rising fast. The cost in 2002 was RMB 68.2 billion; in 2003, RMB 96.6 billion, an increase of 29 percent. However, because of the suppression of interest rates and the use of bank debt (which becomes nonperforming loans) to cover the cost of government programs off-budget, the real costs are far higher.

Local Government Debt and Arrears. The debt problems of local governments can in principle be ameliorated by installing a system of disciplined local borrowing. However, because local governments are not meeting recurrent obligations and have been bailed out by the national government in the past, it is more appropriate to consider local arrears as part of total national debt that must eventually be paid.

Unfunded Pension Liabilities. In addition to these liabilities, the national government will probably choose to continue to make up pension payment shortfalls resulting from the current system. Although it is not legally liable for most current pensions because they are the obligation of the state-owned enterprises, not the Chinese state, for political reasons the government will not be able to avoid these obligations. Pensioners will protest if pensions are not paid. Because of the sensitivity of this issue, protests could easily reach a size to threaten the government. The pension shortfalls are substantial. In 1999, state-owned enterprises owed the pension fund RMB 38 billion. Provincial governments were also coming up short; the government was forced to transfer RMB 17 billion to provinces to cover pension shortfalls in that year, adding a little over 1 percent to total government spending.

---

[33] Based on interviews with officials of international financial institutions.

Assets

The government hopes that it can defray many of its liabilities by selling off some of its assets (Table 3.4). Its principal assets are the SOEs, whose net assets the government judged to be RMB 9.3 trillion in 2001. However, the vast majority of the SOEs have proven to have little or no value. One can count against government liabilities only the government assets the government will reasonably be able to sell in the foreseeable future to cover those liabilities.

The government has already managed to list many of its largest and best enterprises on the stock exchange, typically selling about a 30 percent share of each while keeping majority ownership for itself. At its peak, the stock market reached a capitalization of over 50 percent of GDP, of which somewhat less than a third were traded stocks that had actually been sold to investors. These shares were sold at inflated prices, usually multiples of 50 or so times earnings, or more than three times the historically typical prices of high-quality companies on the New York Stock Exchange. Not to put too fine a point on it, for a while the government managed to sell low-quality companies for several times the prices that the world's best companies command elsewhere. Its ability to do this resulted from the lack of alternatives available to Chinese investors and from the naiveté of those investors. Naiveté has lessened substantially following a long bear market.

Table 3.4
Total Assets of State-Owned Enterprises

	National		Central		Local	
Year	Consolidated Number of Enterprises (1000s)	Total Assets (RMB billions)	Consolidated Number of Enterprises (1000s)	Total Assets (RMB billions)	Consolidated Number of Enterprises (1000s)	Total Assets (RMB billions)
1997	262	12497.5	26	4862.4	236	7635.1
1998	238	13478.0	23	5166.9	215	8311.1
1999	217	14528.8	22	5735.2	195	8793.6
2000	191	16006.8	15	6745.8	176	9261.0
2001	174	16671.0	17	7321.1	157	9349.9

SOURCE: *China Finance Yearbook 2002.*

The government had hoped to fund the social security system by selling shares in a wide range of government-owned companies. However, that announced policy elicited fears of oversupply of stock and caused a large decline in the stock market, so the government backed off. The government has delayed allowing unrestricted listings of private companies in the hope that it will someday be able once again to sell state shares at inflated prices. It has also opened the door slightly to foreign investors in the hope of hawking some state enterprise shares to foreigners.[34] However, its ability to unload its inefficient state sector profitably has somewhat diminished in response to Abraham Lincoln's dictum: "You can fool some of the people all the time, and all of the people some of the time, but not all of the people all of the time."

The IMF has just begun a study to quantify the potential sales value of state-owned assets. The government estimates its assets at about 111 percent of GDP,[35] but many of those assets have no market value whatsoever (RMB 2.1 trillion out of the year 2000 total of RMB 9.9 trillion are government administrative institution assets). Others the government would never sell, and many appear to be valued well above their market value. Looking at Table 3.5, which shows the components of SOE assets, a skeptic might well figure that long-term investments, deferred assets, and other assets might actually be worth zero and that fixed and current assets are probably valued at cost,[36] which for SOEs is often a substantial multiple of market value. When one makes this skeptical calculation and nets the result out against RMB 6.4 trillion of SOE liabilities (also from *China Finance Yearbook* and likely to be a more durable number), one can even make an argument that net SOE assets are negligible.

---

[34] Share prices on overseas markets set a benchmark for the fair value of Chinese enterprises, and even that benchmark is sometimes inflated by China fever. Stock prices for the same or similar companies on the Chinese domestic exchanges run several times higher than the market benchmarks. This is because domestic investors have few investment alternatives.

[35] Dividing the SOE asset numbers in the table above by GDP, both from *China Finance Yearbook 2002*.

[36] In some cases, SOE fixed assets consist mainly of piles of rust but are valued at the original cost of very expensive machinery. More often, the machines are still working but have not been depreciated and may be obsolete.

Table 3.5
Main Assets of State-Owned Enterprises (RMB billions)

Item	1997	1998	1999	2000	2001
Current assets	5369.9	5575.1	5935.2	6682.6	6678.6
Long-term investment	912.0	1097.8	1199.5	1267.9	1366.2
Fixed assets	5787.4	6300.8	6855.3	7466.4	7983.5
Intangible assets	157.3	218.2	243.4	311.3	379.6
Deferred assets	173.0	184.3	190.5	173.9	157.5
Other assets	96.6	101.8	103.7	100.5	100.8
Deferred tax (debt)	1.3		1.2	4.4	4.9
Total assets	12497.5	13478.0	14528.8	16006.8	16670.9

NOTE: 1998 "Other assets" were clearly misstated in the original text by two orders of magnitude, and the "Total assets" figure, while consistent with other years, was less than the sum of its components by a similar amount. We have therefore assumed that the total was correct and have adjusted "Other assets" from 16,006.8 to 101.8.

In any case, the correct number to use for comparison with government debt is the government's equity, not gross assets,[37] and the Chinese government calculates its total equity in the SOEs at RMB 2.9 trillion (see Table 3.6). On a purely market-based calculation, the number would probably be far smaller.

How does one balance these unreliable numbers? Sober observers guesstimate that Beijing might ultimately be able to sell assets, SOEs and others, for as much as 30 to 40 percent of GDP.[38] Based on the calculation in Table 3.6, that could be a huge overestimate. Even that figure assumes that the government is willing to sell off virtually all its useful assets and become one of the least socialist countries on the planet. That may well happen, but for now it is a very strong assumption. We believe 15 percent of GDP is a much better rough estimate,

[37] Shuanglin Lin makes the error of using gross assets—about as big an error as one can make.

[38] Based on conversations with Pieter Bottelier, former head of the World Bank office in Beijing, subsequently a researcher on China at World Bank headquarters, now on the faculty of the Johns Hopkins School of Advanced International Studies. That figure presumes, however, a willingness to sell all state enterprises, a policy not even the United States would contemplate. It is imaginable that over a period of decades China could eventually sell down its state enterprises to a level comparable with, or even below, that of the United States, but it is not imaginable that China would go to zero state ownership.

Table 3.6
Total Owner's Equity of State-Owned Enterprises

	National		Central		Local	
Year	Consolidated Number of Enterprises (1000s)	Total Owners Equity (RMB billions)	Consolidated Number of Enterprises (1000s)	Total Owners Equity (RMB billions)	Consolidated Number of Enterprises (1000s)	Total Owners Equity (RMB billions)
1997	262	4616.5	26	2223.3	236	2393.2
1998	238	5037.1	23	2377.4	215	2659.6
1999	217	5381.3	22	2610.9	195	2770.4
2000	191	5797.6	15	2943.2	176	2854.3
2001	174	6143.6	17	3219.0	157	2924.6

SOURCE: *China Finance Yearbook 2002.*

leaving a very large gap between liabilities and assets. The government can certainly sell more stock in SOEs on stock exchanges, but it will incur serious hidden costs that must be netted out.

China's Government Balance Sheet

In short, China appears to be carrying total government debt of 90–110 percent of GDP, as shown on Table 3.7.

These obligations will have to be covered by the government in the coming years as loan losses are recognized, the state banks are re-capitalized, and the central government absorbs the obligations of insolvent local governments. Against these obligations the Chinese government has salable state assets of perhaps 15 percent of GDP. It also has foreign exchange reserves of roughly one-third of GDP, of which arguably one-third (10 percent of GDP) could be used to cover a fiscal problem in a crunch.

Broader Conclusions

China's government is enjoying the fruits of rapid economic growth. If reform continues to be vigorous, the country is likely to continue

Table 3.7
Total Estimated Chinese Government Liabilities

Liability	Percentage of GDP
Overt government debt	16
International financial institutions bank, including recapitalization bonds	9
Pension transition costs	15
Local government debt	9
Nonperforming loans at 80% of face value	40–60
Total	90–110

SOURCE: Author's analysis.

to reap the attendant budgetary fruits. These fruits are substantial and should include the kind of increase of prestige and influence that other Asian miracle economies enjoyed during their boom years.

However, Chinese policy reflects a dualism, a kind of schizophrenia that will have to be resolved in coming years. On one hand, the Chinese government has been trying to limit state enterprise losses, to stave off a national banking catastrophe, to create substitutes for the health care and pensions and education that the state enterprises will no longer provide, and to rescue an environment increasingly defined by creeping desertification and dangerous annual floods and droughts. On the other hand, China has been acting with the hubris of a secure billionaire, building some of the world's finest buildings, becoming the principal client of the world's great architectural firms, planning to put men in space and then on the moon, constructing the world's largest dam, and diverting one of the world's greatest rivers.

Over the longer term, China will have to enter a new era of sobriety, when total government expenditure (on-budget plus off-budget) will be able to grow only in line with GDP growth. But if rapid growth is to be sustained, the bills have to be paid. Growth in government revenues will have to be used to bail out the state-owned banks, rescue the environment, support the unemployed, cushion the transformation of agriculture, and settle 200 million or so rural-urban migrants. These demands will constrain growth in other budget categories in the coming years, including defense.

CHAPTER FOUR

China's Military Expenditures

Introduction

Chinese military budgets have posed a daunting challenge for students of the People's Liberation Army. Previous analyses, constrained by a lack of empirically robust data, focused on defining the categories of the officially published budget amount, sometimes offering tentative projections of the unknown complete expenditure level. Using newly available primary Chinese-language sources, this chapter expands the state of knowledge on the structure and process of the Chinese military budget system, identifying the key organs involved and the sequence of budget programming. The chapter also tackles the data issue, exploring potential new pools of Chinese data for insights about the missing pieces of the total budget. It improves the state of knowledge of Chinese military budgets by outlining the structure and process of PLA budget programming and employing new empirical sources to better estimate the total financial resources provided the military.

Why We Should Care

Although Chinese military budget figures are often misused as an analytical shortcut for assessing capabilities and intentions, the exercise of disaggregating budgets and tracking them over time is still useful. Military budgets are a visible manifestation of national strategic intentions, priorities, and policies. In this respect, trends in the budget as a percentage of GDP are important, as is the long-term sustainability of spending. The relative scale and dynamism of spending are also a reflection of the state of civil-military relations. Finally, the breakdown of the numbers provides insights into internal military spending and modernization priorities. Properly used, the numbers may even be employed as one supplementary metric of overall military capabilities.

Yet there are clearly limits to the analysis of PLA budgets. First, the PLA is still a relatively opaque institution embedded within a larger system that continues to regard innocuous information like water table data as a national secret. In response to international calls for greater openness, one Chinese military official countered that "transparency is a tool of the strong to be used against the weak."[1] Second, the official PLA budget, as we discuss in greater detail in a later section, does not tell the whole story. China does not conform to international statistical standards for reporting defense expenditures, and major categories are hidden in other parts of the state budget. As a result, a number of outside analyses of PLA budgets fall into the trap of "garbage in, garbage out," extrapolating analytically dubious insights from incomplete, flawed data. Despite these substantial obstacles, however, PLA budgets are a useful source of information on the resources allocated to the Chinese military.

Approach

Chinese expenditures on the military, like Soviet expenditures in the past, are very difficult to track. In addition to the official budget, a number of ministries other than the Ministry of Defense are believed and in some cases are reported to provide additional funds and subsidies. Provincial and local governments also chip in through various programs, including paying conscripts to work on local construction projects.

Previous attempts to assess the total level of Chinese military spending have also been impeded by a lack of understanding of the structure and process of Chinese military budgeting process and the paucity of empirical data.[2] This chapter attempts to correct these de-

---

[1] Interview with PRC defense attaché, Washington, D.C., 2001.

[2] Recent Western writings on Chinese defense expenditures include David Shambaugh, *Modernizing China's Military: Progress, Problems, and Prospects,* Berkeley: University of California Press, 2003; Wang Shaoguang, "The Military Expenditure of China, 1989–98," *SIPRI Yearbook 2000,* Oxford: Oxford University Press, 2000; Wang Shaoguang, "Estimating China's Defense Expenditure: Some Evidence From Chinese Sources," *The China Quarterly,* No. 147, September 1996; Bates Gill, "Chinese Defense Procurement Spending: Determining Intentions and Capabilities," in James R. Lilley and David Shambaugh, eds., *China's Military Faces the Future,* Washington, D.C.: American Enterprise Institute, 1999; Arthur Ding, "China Defense Finance: Content,

ficiencies with newly available Chinese-language primary sources. On the structure and process front, Chinese military authors have published a large literature on the defense budget over the past two decades, offering unprecedented levels of detail on the organizational responsibilities for budget programming, implementation, and auditing, as well as the annual timeline and requirements for programming.[3] A particularly notable official source on budgeting is the *Practical Encyclopedia of Chinese Military Finance,* published in 1993.[4] In addition, Chinese ministerial and provincial statistical data is now widely avail-

---

Process and Administration," *The China Quarterly,* No. 146, June 1996; International Institute for Strategic Studies (IISS), "China's Military Expenditures," *The Military Balance 1995/96,* London: IISS, 1995, pp. 270–75; David Shambaugh, "Wealth in Search of Power: The Chinese Military Budget and Revenue Base," paper delivered to the Conference on Chinese Economic Reform and Defense Policy, Hong Kong, July 1994; and Richard A. Bitzinger and Chong-Pin Lin, *The Defense Budget of the People's Republic of China,* Washington, D.C.: Defense Budget Project, 1994.

[3] Chinese sources on military budgets fall into two categories: journals and books. The journals include *Junshi jingji yanjiu* [Military Economics Research], published by the Military Economics Research Institute in Wuhan, as well as *Jundui caiwu* [Military Finance]. The most important books include Ku Guisheng and Quan Linyuan, *Junfeilun* [On Military Budgets], Beijing: National Defense University, 1999; Contemporary China Series Editing Group, ed., *Dangdai Zhongguo jundui de houqin gongzuo* [Contemporary Chinese Military Logistics Work], Beijing: Zhongguo shehui kexue chubanshe, 1990; Zhang Xulong, ed., *Junshi jingjixue* [Military Economic Science], Shenyang: Liaoning renmin chubanshe, 1988; Long Youcai and Wang Zong, eds., *Jundui caiwu jianshe* [Military Economic Construction], Beijing: Jiefangjun chubanshe, 1996; Wang Qincheng and Li Zuguo, eds., *Caiwu daquan* [Finance Encyclopedia], Urumqi: Xinjiang renmin chubanshe, 1993; Lu Zhuhao, ed., *Zhongguo junshi jingfei guanli* [China's Military Budget Management], Beijing: Jiefangjun chubanshe, 1995; National Defense University Development Institute, ed., *Zhongguo guofang jingji fazhan zhanlue yanjiu* [Research on China's National Defense Economic Development Strategy], Beijing: Guofang daxue chubanshe, 1990; Sun Zhenyuan, *Zhongguo guofang jingji jianshi* [China's National Defense Economic Construction], Beijing: Junshi kexueyuan chubanshe, 1991; Gao Dianzhi, *Zhongguo guofang jingji guanli yanjiu* [Research on Chinese Defense Economic Management], Beijing: Junshi kexueyuan chubanshe, 1991; Jin Songde, et al., *guofang jingji lun* {On National Defense Economics], Beijing: Jiefangjun chubanshe, 1987; Zhang Zhenlong, ed., *Junshi jingjixue* [The Science of Military Economics], Shenyang: Liaoning renmin chubanshe, 1988; Chinese Military Encyclopedia Editing Group, ed., *Jundui houqin fence* [Military Logistics Volume], Beijing: Junshi kexueyuan chubanshe, 1985; People's Liberation Army Logistics College Technology Research Section, ed., *Junshi houqin cidian* [Military Logistics Dictionary], Beijing: Jiefangjun chubanshe, 1991; and Lin Yichang and Wu Xizhi, *Guofang jingjixue jichu* [Basic Defense Economics], Beijing: Junshi kexueyuan chubanshe, 1991.

[4] China Academy of Military Sciences Editing Group, ed., *Zhongguo junshi caiwu shiyong daquan* [Practical Encyclopedia of Chinese Military Finance] (hereafter , ZJCSD), 1993.

able to Western researchers, permitting new empirical inquiries into the previously opaque off-budget spending categories.

Organization of This Chapter

Following this introduction, the chapter is divided into five main sections. The first outlines the structure and process of Chinese military budgeting, focusing on the organizations involved in preparing and implementing the annual budget as well as the sequence of the budget programming itself. Using Chinese numbers, the second section analyzes the official budget, teasing out as many solid conclusions as possible from the artificially constructed figures. The third section takes a close look at extra-budgetary revenue, including an examination of revenues derived from commercial companies owned and operated by the PLA, and the impact of divestiture on the PLA and, in particular, PLA finances. The final section attempts to extrapolate from the official budget to the total budget.

Structure and Process of Chinese Military Budgeting

Introduction

Previous forays into the morass of Chinese military budgeting have, with one or two notable exceptions,[5] tackled the numbers themselves rather than the process that generates the numbers. There are three reasons for changing course and examining the system itself. First, the numbers-driven effort has run aground in the shallow water of official Chinese statistics. Second, Chinese-language sources are now available that explicate the budget process. Third, and most important, an examination of the structure and process of Chinese military budgeting is essential to help the analytical field dislodge itself from its current predicament, because a better understanding of the programming cycle and its components will likely point the way to new pools of empirical data.

---

[5] See Shambaugh, 2003, Chapter 5.

Key Budget Organizations[6]

The overall PLA budget organization system is summarized in Figure 4.1 below.

At the top of the system, the Party leadership in the Politburo Standing Committee, the rump Politburo, and the Central Committee set overall strategic guidelines and direction for the country, including the importance of military funding relative to other national priorities, such as economic modernization. On the right side of the figure are the civilian governmental organs, headed by the State Council un-

Figure 4.1
The PLA Budgeting Organizational System

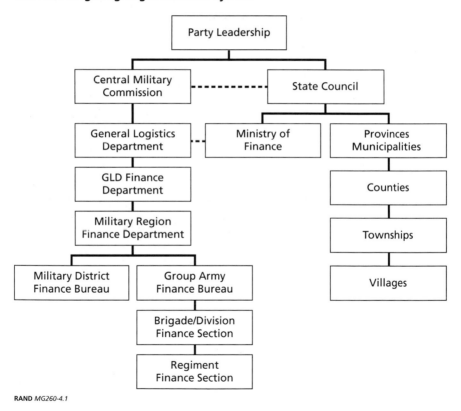

RAND *MG260-4.1*

---

[6] The explanation of budget organizations in this section comes from ZJCSD, 2003, pp. 137–142.

der the leadership of China's prime minister. While many government bodies have responsibilities for oversight and management of parts of the financial and economic system (collectively known as the "national finance apparatus," or *guojia caizheng*), the Ministry of Finance has "budget responsibility" *(yusuan baogan)* for developing the national budget, and "leads and administrates . . . defense expenditure and assets" (known as national defense fund expenditure quotas, or *guofang zhichu zhibiao*).[7] Below Beijing, civilian government units at the provincial, municipal, county, township, and village level have responsibilities for providing support monies to the military units and organizations within their jurisdictions. On the military side of the chart, the Central Military Commission (CMC) is the top bureaucratic organ. It "manages defense outlays and assets jointly with the State Council," including the top-line expenditure level for a given year as well as the projections contained in national five- and ten-year plans.[8] The General Logistics Department's (GLD's) Finance Department is "the CMC's highest money management organ"[9] and the "army's functional department" *(jundui zhineng bumen)* for money management *(licai)*. It is responsible for

- organizing and guiding the Army's implementation of the Party and government's financial policies
- formulating the military's financial laws and regulations
- producing the total annual military budget and final accounting for civilian ministries
- organizing and overseeing military accounting work
- guiding economic production work
- managing funds for "strategic material stores" *(zhanlue wuzi chubei)* and overseeing "circulating funds" *(zhouzhuanjin)*
- setting military industrial product prices
- organizing wartime finance work

---

[7] *China's National Defense in 1998.*

[8] *China's National Defense in 1998.*

[9] ZJCSD, p. 137.

- organizing financial personnel training and evaluating technologies for financial work
- supervising financial investigations of lower-level units.

Within the PLA, the GLD budget coordination process is facilitated by a vertical hierarchy of GLD-subordinate financial bureaus and sections at every level of the system from the military region down to the regiment. At the same time, however, significant portions of the PLA lie outside of the GLD's purview. Eight central-level PLA institutions, including the General Staff Department, General Logistics Department, General Political Department, General Equipment Department, Ministry of Defense, Commission on Science, Technology, and Industry for National Defense, the Strategic Rocket Forces, and the universities and colleges of the professional military education system make their annual budget bids directly to the Central Military Commission, bypassing the process governing the geography-based administrative units and combat units.

The Military Budgeting Process[10]
The Chinese military budgeting process provides an important window into PLA institutional interests and priorities. Interestingly, it is also one of the few areas of regular, standardized interaction with the rest of the Chinese bureaucracy, and thus offers unique insights into the military's role and tactics within the system.

At a macro level, Chinese sources reveal that there are actually *three* military budget processes, all operating simultaneously. The first, known as the centralized *(tongguan)* process, has predominated, and involves direct expenditures from central government coffers on national, regional, and district-level military units and organizations. The second, referred to as the "decentralized" *(fenguan)* process, supplements the centralized process with funds contributed by civilian government units at all levels. This system is also known as the "three-thirds" *(san fenzhi san zhidu)* system, since military units at the central department,

---

[10] This section draws on Shambaugh, 2003, pp. 205–210.

military region, and district levels receive allocations from provincial/ municipal, county, and local governments. A third system, relevant to certain budgetary items, consists of a combination of the centralized and decentralized processes.

The primary centralized system uses what the Chinese call a "down-up-down" *(zishang erxia)* dynamic, whereby national-level civilian and military units (led by the Ministry of Finance and the Central Military Commission, respectively) establish top-line budget targets, then military regions and district units bid for funds in a sequentially aggregated process from the lowest level to the highest level, and the center finally sets expenditure ceilings and transmits the results of the bidding to the lowest levels.

The annual budget cycle can be broken down according to the Chinese fiscal calendar, which begins each March with the announcement of the central budget by the Minister of Finance at the National People's Congress (see Figure 4.2). In accordance with the programming of the relevant five-year and ten-year plans as well as nearer-term changes made by the central leadership, the "national finance" *(guijia caizheng)* apparatus under the State Council agrees on a top-line expenditure number (known as the "military allocation plan" or *junfei bokuan jihua)* at the beginning of April and transmits this number to the Central Military Commission (CMC), which is responsible for "overall planning for all the military's expenditures." One PLA finance source complains that this number is sometimes set relatively late, which "is not good for army planning." Under CMC guidance, the General Logistics Department's Finance Bureaus down to the division level transmit the new annual funding targets *(jingfei zhibiao)* and "calculate the needs" *(xuyao liang)* of the army by assembling bids from ground, navy, and air force units.

These proposals, which are called "investigation and argumentation reports" *(diaocha lunzheng baogao)*, are passed to the military district logistics departments in July, and on to the military region logistics departments in the August–September timeframe. The budget draft reports themselves use a "rolling calculation method" *(gundong jisuan fangfa)*, in which the previous year's allocation is the "base number" *(jishu)* that is adjusted with "unfunded expenditures" *(zhuanjiaxing zhichu)* and "newly increased needs" *(xinzeng xuyao e)* in budget cat-

Figure 4.2
PLA Budget Calendar

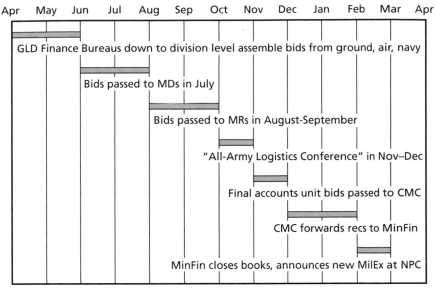

RAND *MG260-4.2*

egories. At each level, the draft report is submitted to the relevant party committee for "examination and approval" *(shenpi tongyi)*. An "All-Army Logistics Conference" is traditionally held sometime in November or December. The GLD Finance Department then takes the total budget draft and submits it for "investigation and consent" *(shencha tongyi)* to the GLD headquarters, which then transmits it to the Central Military Commission for higher-level "examination and approval" *(shenpi tongyi)*. In January, the Central Military Commission develops the recommended budget draft *(jingfei yusuan caoan)*, which includes three parts: (1) budget statement *(yusuan biao)*; (2) budget plan explanation *(yusuan anpai shuoming)*; and (3) measures for implementing the budget *(zhixing yusuan cuoshi)*. The draft budget package is then distributed to "related departments" *(youguan bumen)* for comments, and then the "total annual military budget" *(junfei niandu zongyusuan)* is forwarded for final deliberation to the Ministry of Finance, which closes the books in February and announces the new military expenditure number at the March National People's Congress meeting. If there is a surplus, the national financial authorities are not notified, and the monies (known as "the balance of financial resources" or *junfei zong jiesuan*) can be spent under the principle of "annual continuous usage *(niandu jixuu shiyong)*. If there is overspending, the national financial authorities do not adjust the allocation *(buzai bubo)*. Instead, responsibility for the deficit lies with the Central Military Commission.

Changes in the Military Budgeting Process

In recent years, important changes have occurred in the defense budget process. *China's National Defense in 2002* (the *2002 Defense White Paper)* asserts that "the administration of defense expenditure has undergone a whole array of reforms, including reform in defense expenditure budget methods, centralized payments for weapons and equipment procurement, and a tendering and bidding system for the procurement of defense materials, projects and services."[11] In March 2001, reforms of the budget planning system were initiated, in particular the implementation of a zero-base budgeting method. As noted above, prior to this reform, units and administrative organs began budgeting from the assumption that the previous year's allocation would be used as the base for the next year's planning. In addition, they also assumed they would be permitted to roll over any surpluses or unspent monies from the previous year into the next. The new system required them to rack and stack their proposed budget from zero every year. In January 2002, changes were made in the procurement system. Contracts for bulk materials exceeding RMB 500,000 ($60,000) in value or construction projects exceeding RMB 2 million ($240,000) in value were subject to public bidding. The goal of these changes is a "more just, fair and transparent" defense funding system.[12]

---

[11] *China's National Defense in 2002.*

[12] *China's National Defense in 2002.*

The Official PLA Budget: What Can It Tell Us?

What Do We Know? What Don't We Know?[13]

Our knowledge of Chinese military budgets relies heavily on official information provided by the government in Beijing. As a consequence, the universe of available information is small and highly aggregated in nature. From open sources, we know

- the official "top line" of PLA expenditures
- the official budget as a percentage of government spending and GDP
- a rough breakdown of official defense expenditures by large category.

The first two are derived from Chinese statistical yearbooks, while the latter has appeared in China's *1998, 2000,* and *2002 Defense White Papers.*

Not surprisingly, the list of what we do not know is much longer, and includes even highly aggregated data:

- The "real" top-line budget
- Contributions to the PLA by civilian government units
- Extra-budgetary *(yusuanwai)* expenditures
- Expenditures by service branch
- Expenditures by PLA budget functional category
- Longitudinal numbers of all the above.

The task is further complicated by the fact that the true budget of the PLA is so disaggregated within different Chinese ministry budgets that the Chinese military itself may not even know the bottom-line figure.

---

[13] This section draws on Bitzinger, 2003, pp. 167–169.

Definitions: What Does the Official Budget Include? What Is Excluded?

As described in multiple official Chinese sources, in particular the *1998 Defense White Paper*, the official PLA budget includes three main categories:

- **Personnel** *(shenghuo fei)*: "pay, food, clothing for military and nonmilitary personnel"
- **Operations and Maintenance** *(shiye he gongwu fei)*: "training, construction, maintenance of facilities, and operating expenses, education and combat costs"
- **Equipment** *(zhuangbei fei)*: "costs for equipment, including research and experimentation, procurement, maintenance, transportation and storage."

The *1998 Defense White Paper* adds: "In terms of the scope of logistics support, these expenditures cover not only active service personnel, but also militia and research requirements. In addition, a large amount of spending is used to fund activities associated with social welfare, mainly pensions for retired officers, schools and kindergartens for children of military personnel, training personnel competent for both military and civilian services, supporting national economic construction, and participating in emergency rescue and disaster relief efforts."

Internal Chinese military sources provide more detailed insights into the key components of the overall PLA budget, which can be divided into fifteen functional categories:

- Living expenses *(shenghuo fei)*
- Official expenses *(gongwu fei)*
- Operating expenses *(shiye fei)*
- Education and training expenses *(jiaoyu xunlian fei)*
- Equipment procurement expenses *(zhuangbei gouzhi fei)*
- Logistics procurement and maintenance expenses (houqin zhuangbeu gouzi weixu fei)

- Weapons maintenance management expenses *(zhuangbei weichi guanli fei)*
- Scientific research expenses *(kexue yanjiu fei)*
- Militia expenses *(minbing fei)*
- Fuel expenses *(youliao fei)*
- War preparation and combat expenses *(zhanbei zuozhan fei)*
- Miscellaneous and flexible expenses *(qita he jidong jingfei)*
- Central Military Commission reserve funds *(Junwei dingwu fei)*
- Basic construction expenses *(jiben jianshe fei)*
- People's Air Defense expenses *(renmin fankong fei)*.

However, these sources also make clear that these functional categories are not the same as the contents of the "official" Chinese defense budget, announced each March by the Minister of Finance at the National People's Congress.

The official budget excludes a wide variety of military accounting items commonly included in Western budgets:

- Procurement of weapons from abroad
- Expenses for paramilitaries (People's Armed Police)
- Nuclear weapons and strategic rocket programs
- State subsidies for the defense-industrial complex
- Some defense-related research and development
- Extra-budget revenue *(yusuanwai)*.

These budget items are included in other parts of the state budget. Confirmation of this can be found in the *2002 Defense White Paper,* which plainly states that "the entire defense expenditure comes from the state financial budget."[14] The funds for foreign weapons procurement, which one U.S. government source asserts averaged $775 million through much of the 1990s and is currently estimated at $3

---

[14] *China's National Defense in 2002.*

billion per year,[15] are reportedly drawn from specially-arranged hard-currency accounts controlled by the State Council, not the Chinese military. Money for paramilitaries is divided among the Ministry of Public Security, ministries employing People's Armed Police (PAP) personnel, and localities with PAP units. While retirements and pensions are provided only to officers of senior rank, the remainder of the force receives a one-time demobilization payout. Additional monies come to the PLA from local government funds, under the aegis of the "three-thirds" principle, established by the Central Military Commission in the 1990s, whereby local governments and internal sources of revenue are sources of supplemental funding.[16]

The Official Budget Numbers

The Chinese official defense budget from 1978 to the present in nominal renminbi is shown in Figure 4.3. To give a sense of the inflation-

Figure 4.3

Chinese Official Defense Budget, 1978–2003

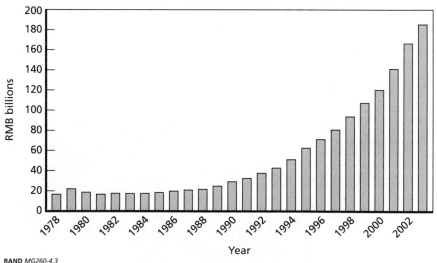

---

[15] U.S. Department of State, 2000, p. 129.

[16] Lu Zhuhao, p. 344.

adjusted value of these expenditures, we have converted these figures into 2002 renminbi in Figure 4.4 using China's GDP deflator.

Since 1978, the budget has grown 11-fold, from RMB 16.8 billion to more than RMB 185.3 billion in 2003.[17] The *2002 Defense White Paper* outlines five reasons for the significant increases in military spending, particularly since 1989:

1. Increase of personnel expenses. Along with the socioeconomic development and the rise in per-capita incomes of rural and urban residents, it is necessary to improve the living standards and conditions of military personnel. The past decade has witnessed the increase in compensation for living expenses in the armed forces on five occasions, and an 84 percent increase in salaries for officers and a 92 percent increase in allowances for soldiers.

Figure 4.4
Official Defense Budget in Constant 2002 Renminbi, 1978–2004

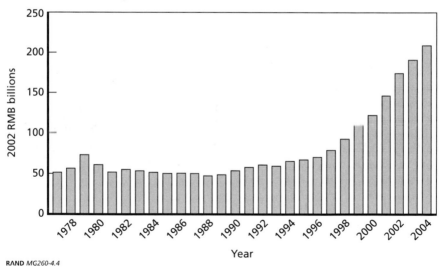

---

[17] During the same period, GDP grew 28-fold and central government budgets grew almost 17-fold.

2. Establishment and gradual improvement of a social security system for servicemen. In accordance with the requirements of the market economy, China has, since 1998, established such systems as disability and life insurance for servicemen, medical insurance for demobilized servicemen, and housing subsidies for servicemen, and has adjusted and enhanced living expense standards for PLA retirees.
3. Increased expenses for operations. Since the armed forces stopped commercial activities in 1998, the budget for training has increased year by year. With the gradual improvement of living facilities and progressive office automation, expenses of a maintenance nature have kept rising.
4. Increased expenses for cooperation with the international community in antiterrorism activities.
5. Increased expenses for the renovation and improvement of military equipment through technological improvements, particularly through the use of more sophisticated technologies.

Not mentioned, but equally important, are the political and civil-military imperatives of keeping the military happy, as well as the strategic imperatives of a potential Taiwan contingency involving U.S. military intervention.

Minister of Finance Xiang Huaicheng announced on March 6, 2003, that defense spending in 2003 would rise 9.6 percent to RMB 185.3 billion ($22.4 billion). The official reasons for the increase included "changes in the international situation, safeguarding China's national security, sovereignty and territorial integrity, and raising the combat effectiveness of the armed forces in fighting wars using high technology."[18] Yet a 9.6 percent increase represented a significant drop-off in the rate of growth of the PLA budget. Indeed, the 2003 budget represented the first time in 14 years that the PLA did not receive a double-digit year-on-year increase. One official source offered a reason for the smaller-than-normal increase, arguing that slower overall

---

[18] Xiang Huaicheng at NPC, "PRC Plans 9.6 Percent Increase in 2003 Defense Spending," *Xinhua,* 6 March 2003.

economic growth required caps on central budget spending.[19] A hint of another reason can be found in the fact that only official English-language sources, such as *China Daily*, highlighted the drop in the rate of increase as the "lowest in 14 years"[20] whereas Chinese language sources merely stated the numbers without editorial comment. The 2003 increase was well below the projected programming of the tenth Five Year Plan, which appeared to be averaging between 15 and 20 percent after inflation.

What is going on here? Although the official budget numbers were already widely viewed as incomplete, it is entirely possible that the Chinese government, weary of the annual public relations debacle in the Western media over double-digit increases in its defense budget, decided to hide a greater share of the increase in other accounts in 2003. Using this logic, 9.6 percent was a reasonable compromise between previous high-profile increases of nearly 18 percent and lower amounts, such as 5 percent, that would have been politically embarrassing to the important military constituency.

Nonetheless, numerous PLA officers publicly called the increase insufficient and argued for greater resources. People's Liberation Army Air Force (PLAAF) Lieutenant General (LTG) Liu Cangzi allegedly told the *South China Morning Post* that defense spending should be increased "many, many times," while his colleague LTG Zeng Jianguo told the same paper that the budget should be raised "even more in certain respects."[21] Even more shocking were the comments of Major General Ding Jiye, the head of the General Logistics Department Finance Department, who told the state-run Xinhua news agency that the current level of defense spending was "barely enough to keep things moving."[22] One PLA delegate asserted that the level of military mod-

---

[19] Allen T. Chung and Fong Tak-ho, "Defense Outlay Sees Smallest Rise in 14 Years," *South China Morning Post*, 7 March 2003.

[20] Xing Zhigang, "Military Budget Rise Lowest in 14 Years," *China Daily*, 7 March 2003.

[21] Allen T. Chung and Fong Tak-ho, "Defense Outlay Sees Smallest Rise in 14 Years," *South China Morning Post*, 7 March 2003.

[22] "Major General Ding Jiye Says Defense Spending Much Lower Than World Average," *Xinhua (English)*, 8 March 2003.

ernization is only "on par" with capabilities of major countries in the 1970s and is "incompatible" with China's "comprehensive national strength" 20 years after reform.[23] To correct these deficiencies, delegates called for the leadership to "raise the welfare and remuneration of military officers and men, improve the living conditions of military officers on active duty, increase allowances for officers and men on active duty, and narrow the gap between military personnel on active duty and other civic servants in terms of welfare and wages."[24]

More important than the absolute increases themselves, however, Figure 4.5 shows that PLA budgets have only outpaced inflation in significant measure since 1993, after declining in real terms in the 1980s and during the inflationary cycles in the late 1980s and mid-1990s.

After 1996, the double-digit increases finally took hold, resulting in net gains for military expenditure in real terms. In more recent years,

Figure 4.5
Real and Nominal Changes in PLA Budgets, 1978–2004

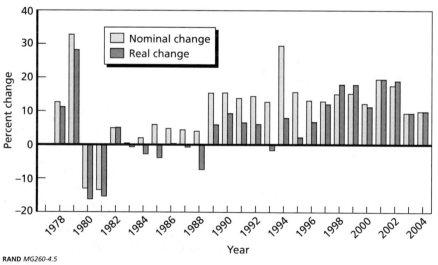

RAND *MG260-4.5*

---

[23] Cao Zhi, "Proper Investment Into National Defense Is Needed for Building Well-Off Society in All-Around Way," Xinhua, 6 March 2003.

[24] Cao Zhi, op. cit.

China has actually suffered from bouts of mild deflation, giving the annual allocation more buying power.

Creative Approaches to the Official Numbers

Properly massaged, even the official Chinese defense budget numbers can reveal important analytical insights. These findings fall into two broad categories: (1) comparisons between PLA budget trends and overall central budget and economic trends, and (2) trends within the PLA budget itself.

Military expenditures as a percentage of central government spending have dropped significantly, as shown in Figure 4.6. They have fallen from a wartime high of more than 40 percent in 1950 to approximately 8 percent in 2002. Even if one only examines spending since the onset of economic reforms in the late 1970s, military expenditures as a share of government expenditures have fallen almost by half, from 15 percent to 8 percent. The big winners since 1978 have been government administration, and social/educational spending, which rose 15 and 14 percentage points, respec-

Figure 4.6
Military Expenditure as a Percentage of Central Government Spending, 1978–2003

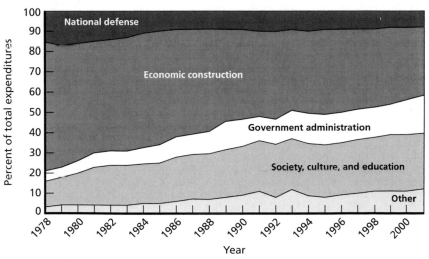

RAND *MG260-4.6*

tively, as a share of government expenditure. As a share of GDP, military spending has also fallen, from 5.5 percent in 1979 to a low of 1.06 percent in 1996; it has since risen to 1.72 percent in 2003 (Figure 4.7).

Other interesting trends can be discerned within the PLA budget itself. Chinese sources reveal that from 1950 to 1980, the ground forces received 50 percent of the budget, compared with 31.37 percent for the air force and 18.4 percent for the navy. Other Chinese sources relate that personnel expenditures from 1950 to 1970 accounted for 40 percent of the overall budget, dropping to 30 percent in the 1970s and rising again to 40 percent in the 1980s. Since 1997, the official budget has been divided roughly equally among personnel, operations and maintenance, and equipment at one-third apiece, based on data on internal breakdowns revealed in the *1998, 2000,* and *2002 Defense White Papers* (Figure 4.8). Compared with other militaries, the PLA spends a relatively smaller amount of its total budget on personnel costs, although the large demobilizations since the early 1980s have undoubtedly suppressed sharp increases in human resource expenditures. In contrast, the Taiwan military spends nearly 55 percent of its defense allocation on personnel.

Figure 4.7
The Official PLA Budget as a Share of GDP, 1978–2003

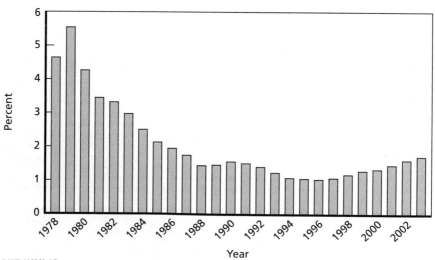

Figure 4.8
Changing Priorities in the Official PLA Budget, 1997–2002

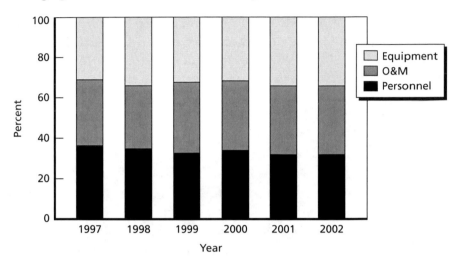

SOURCE: *China's National Defense* in 1998, 2002. Defense White Paper,
1998, 2000, 2002.
RAND *MG260-4.8E*

Extrabudgetary Revenue, PLA Inc., and Divestiture

The issue of extrabudgetary revenue requires a brief discussion of the
Chinese military's historic participation in the economy, the commer-
cialization of the PLA infrastructure in the 1980s and 1990s, and the
subsequent "divestiture" of the military from commercial operations in
the summer of 1998. A distinction must first be made between military economic activities and military resources. The latter, which were
discussed in detail in the previous section, are a function of the official budget and defense expenditures hidden in other parts of the
state budget. The profits and losses from military economic activities,
in contrast, benefited individuals, units, and the PLA as an institution
to varying degrees. Unit enterprises were legally expected to pay taxes
to the national-level authority overseeing the business system, but the
data show that the enterprises generating profits rarely remitted the
full amount of taxes due and that most enterprises were subsistence

operations by design. Units retained most of their revenues in order to bolster local troops standard of living, which provided only indirect benefit to the PLA as a whole. In addition, widespread corruption benefited individual officers or soldiers. Although these individuals were military personnel, the benefits to them did not equate with benefits to the PLA and likely had a net negative effect upon the PLA as a military organization.

The Chinese military has been involved in economic activities since the mid-1920s. Throughout the Mao period, the PLA expanded its internal economic empire, which included large farms, factories, and other units that made the military relatively self-sufficient, at least in terms of food and other supplies for soldiers. After Deng Xiaoping's approval of limited market reforms in the late 1970s, the PLA was authorized to gradually commercialize parts of its economic infrastructure, primarily to generate revenues to compensate for some of the significant reductions in defense budgets that occurred during that period. By the end of the 1980s, the PLA oversaw a vast commercial system of more than 20,000 enterprises.

Both the military and civilian leaderships recognized the problem of military units focusing on commercial activities, but they believed they were trapped in a Catch-22: They could not force the PLA to divest itself of these businesses without replacing the lost revenue with budgetary appropriations (Figure 4.9). However, China's political leaders had little desire to increase military spending to replace revenues that would disappear if commercial activities were to be banned.[25] Their focus was on encouraging economic growth through investments

---

[25] This analysis distinguishes among PLA enterprise revenues (sales), the wage bill, and profits, the latter two equaling value-added. Because wages go to the workers or managers, they may help the PLA retain soldiers and officers but do not provide funds for PLA expenditures. Moreover, the conflict of interest for military personnel intrinsic in being employed by a commercial enterprise while wearing a uniform raises questions about the utility of keeping soldiers and officers in the military who are spending substantial amounts of time on commercial activities. Profits and possibly depreciation allowances could be siphoned off for true military expenditures (procurement, R&D, and operations and maintenance), but the reported net flows from these sources for true military expenditures looks very modest.

Figure 4.9
PLA Business Revenues and Profits, 1978–1996

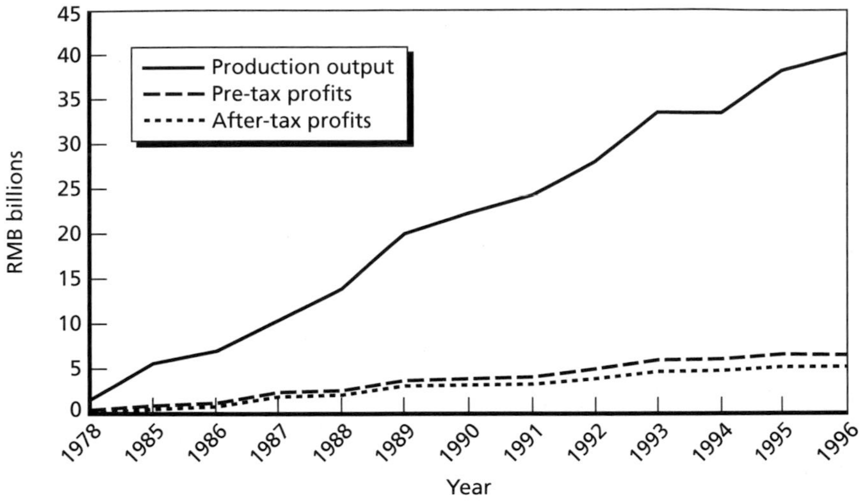

SOURCE: Tai Ming Cheung, 2002.
RAND *MG260-4.9*

in infrastructure. In addition, a number of senior government officials who benefited from the activities of these enterprises had a financial interest in supporting commercial enterprises owned by the military.

Initially, policymakers tried to fix the military commercial system. To this end, throughout the 1990s the PLA conducted a series of increasingly strident rectification campaigns against abuses. Ultimately, these programs were perceived to have failed, culminating in the decision in 1998 to divest the military of its commercial enterprises. On July 22, 1998, at an enlarged session of the Central Military Commission, CMC Chairman Jiang Zemin gave a speech in which he called for the dissolution of the military-business complex, asserting:

> To make concerted efforts to properly develop the army in an all-around manner, the central authorities decided: The army and the armed police [*wu jing*] should earnestly screen and rectify [*qingli*] various commercial companies operated by their subordinate units, and shall not carry out any com-

mercial activities in the future . . . Military and armed police units should resolutely implement the central authorities' resolution and fulfill as soon as possible the requirements that their subordinate units shall not carry out any commercial activities in the future.[26]

Jiang then sought to consolidate the decree by publicizing the announcement through the party's extensive propaganda apparatus. That night, Jiang's speech at the meeting was broadcast on the CCTV *Evening News,* which has the highest rating in China and is closely watched by other Chinese media for cues about important stories. Observers took special note of the fact that the Chinese leader was shown flanked by the top brass of the PLA, projecting at least tacit consent by the military to the decision. The next day, the Party's official newspaper, *People's Daily,* ran a banner headline, declaring "PLA Four General Departments Convened in Beijing to Carry Out the Decision of the Anti-Smuggling Meeting," with the subtitle "Chairman Jiang Talked Seriously About Divestiture."[27] The announcement was then publicly seconded in subsequent days by key members of the military and civilian leadership, including the head of the PLA, General Zhang Wannian, Chief of the General Staff General Fu Quanyou (July 23),[28] General Logistics Department Director Wang Ke (July 24),[29] General Political Department Director General Yu Yongbo (July 25),[30] and General Armament Department Di-

---

[26] "Jiang Orders PLA-Owned Firms to Close," *Xinhua Domestic Service,* July 22, 1998, in Foreign Broadcast Information Service, China Daily Report (hereafter FBIS-CHI), FBIS-CHI-98–204, July 23, 1998.

[27] *People's Daily,* July 23, 1998, p. 1.

[28] "Fu Quanyou on Supporting Jiang's Anti-Smuggling Drive," *Xinhua Domestic Service,* July 23, 1998, in FBIS-CHI-98–206, July 25, 1998.

[29] Cao Haili, "The Chinese Army Has Sailed Out of the Business Sea," *Caijing* [Finance and Economics], January 1999, pp. 1–16.

[30] "Yu Yongbo Calls on Army to Cease Business Operations," *Xinhua Domestic Service,* July 26, 1998, in FBIS-CHI-98–208, July 27, 1998.

rector General Cao Gangchuan (July 26),[31] as well as Politburo Standing Committee member and Jiang Zemin's heir apparent, Hu Jintao.[32]

Contrary to the conflictual civil-military scenario put forward by many observers in the Hong Kong media, interviews in Beijing strongly suggest that divestiture in principle was largely supported by a corruption-weary military leadership.[33] They generally agreed with the political, military, and economic rationales for divestiture. On the political front, divestiture was aimed at curtailing corruption within the ranks, which had been publicly criticized at the very highest levels of the PLA leadership as early as 1993, though never explicitly linked to commercial activity.[34] Painfully aware of the bureaucratic and fiscal Catch-22 created by increasing dependence on proceeds from commercial enterprises owned by the military, the civilian leadership nonetheless summoned the political courage to publicly argue that as long as the military operated in the commercial economy, it was subject to "negative influences." Jiang Zemin even reportedly spoke of preventing the military from "changing color" and of keeping the military "pure." At a military level, divestiture was designed to return the PLA to its primary mission of preparing for war, which was the principal aspiration of an increasingly professional officer corps.[35] Finally, from an economic perspective, there was recognition that the military had never proven adept at running commercial operations and that the failed experiment should end.

Nor did there appear to have been any major cleavages in the top civilian leadership over divestiture.[36] One well-informed observer

---

[31] "General Armament Department to Fight Smuggling," Xinhua Domestic Service, July 26, 1998, in FBIS-CHI-98–209, July 28, 1998.

[32] Wu Hengquan, Liu Zhenying, and Wang Jinfu, "Hu Jintao Speaks on Banning PLA Businesses," *Xinhua Domestic Service,* July 28, 1998, in FBIS-CHI-98–209, July 28, 1998.

[33] Interviews in Beijing with knowledgeable Chinese and Western interlocutors, 1998–2001.

[34] Liu Huaqing and Zhang Zhen, "Carrying Forward Fine Traditions is a Major, Strategic Issue of Our Army's Construction Under the New Situation," *Renmin ribao* [People's Daily], July 26, 1993, pp. 1, 4, in FBIS-CHI-93–142, July 27, 1993, pp. 22–25.

[35] Mulvenon, 1997b.

[36] Willy Wo-Lap Lam, "Problems Between CCP, Army," *South China Morning Post,* October 20, 1999, p. 19.

relates that Jiang and Zhu were closely united on the issue, with Jiang providing the political clout and Zhu providing economic instructions to his subordinates at the State Economic and Trade Commission as to the specifics of the separation. A top-level, civilian-led leading group was quickly established, with Hu Jintao as the head, and other party, government, and military leaders, including Zhang Wannian, and Luo Gan, as members.[37] Hu's appointment served an important prelude to his official appointment as vice-chairman of the Central Military Commission at the end of October 1999.[38] Over the next few weeks, corresponding small leadership groups at lower levels of the system, including military units and State Economic and Trade Commission branches, were also established.

A key condition for military acquiescence to divestiture was an assurance from the civilian leadership that the PLA would receive a sufficiently generous compensation package for handing over its businesses. Indeed, sources in Beijing confirm that the fault lines in the divestiture process could be drawn between supporters, including the senior military leadership and the combat units, and those who resisted the ban, especially members of the logistics and enterprise management structure, military region commands, and military district commands, who stood to lose their primary source of legal and illegal income.[39]

The heart of the bargain between the PLA and the civilian leadership centered on two separate financial deals. The first was the one-time transfer of the PLA's divested enterprises. It is important to note that the divestiture process drew a sharp distinction between "commercial" *(shangye)* enterprises, defined as those enterprises that dealt with customers, and "production" *(shengchan)* enterprises, which includes traditional internal PLA economic units like farms and uniform factories. The former were slated for divestiture; the latter would be retained in some form. Reportedly, the financial burden for the divested enter-

---

[37] "Military Meets Resistance in Enforcing Ban on Business," *Ming pao,* September 9, 1998, p. A16, in FBIS-CHI-98–253, September 10, 1998.

[38] "Hu Jintao Appointed CMC Vice-Chairman," *Xinhua,* October 31, 1999.

[39] Personal communication with Tai Ming Chueng.

prises, including their weighty social welfare costs and debts, was to be placed upon local and provincial governments rather than the central government, although no money was to change hands. This devolution of responsibility from the center to the localities was seen by many as yet another attempt by Zhu Rongji to restore some measure of central authority in China by forcing the lower levels of the system to assume greater financial responsibility for the economic units in their area.

The second negotiation focused on annual budget increases to make up for the additional revenues from enterprise activities lost when the enterprises were divested, with the goal of consolidating Jiang's earlier decree to the military to "eat imperial grain" rather than rely on business for revenue. Before the divestiture was completed, Hong Kong sources reported that the PLA would receive between RMB 15–30 billion per year ($1.8 to $3.6 billion), with the exact time frame subject to negotiation.[40] Two months later, the same author reported that the PLA would receive RMB 50 billion ($6.0 billion) as compensation for its lost enterprises.[41] The *Wall Street Journal* quoted U.S. diplomats as saying the government offered about $1.2 billion, but the military demanded $24 billion.[42] Sources at the GLD claimed in December 1998 that the PLA would receive RMB 4–5 billion ($0.5 to $0.6 billion) in additional annual compensation, complementing continued double-digit budget increases.[43]

For local units, however, the prospects of a lucrative budget deal must have been bittersweet, since it required them to buy into what might be called "the trickle-down theory of PLA economics." Whereas

---

[40] The RMB 15 billion number comes from Kuang Tung-chou, "Premier Promises to Increase Military Funding to Make Up For 'Losses' After Armed Forces Close Down All Its Businesses," *Sing tao jih pao,* July 24, 1998, p. A5, in FBIS-CHI-98–205, July 24, 1998. For the RMB 30 billion figure, see Willy Wo-Lap Lam, "PLA Chief Accepts HK47 Billion Payout," *South China Morning Post,* October 9, 1998.

[41] Willy Wo-Lap Lam, "PLA to Get HK28 Billion for Businesses," *South China Morning Post,* August 3, 1998.

[42] Matt Forney, "A Chinese Puzzle: Unwinding Army Enterprises," *Wall Street Journal,* December 15, 1998.

[43] The author thanks Dennis Blasko for this information.

units previously had relatively direct control over enterprise finances, they now had to place their faith in the notion that the budget funds would trickle down through the system from Beijing to their level. Previous experience with the Chinese military bureaucracy did not inspire confidence that this would come to pass. To ameliorate these concerns and boost morale, the military leadership took steps in the fall of 1998 to improve the standard of living for the rank and file. The principal measure was an increase in the salaries of servicemen by an average of 10 to 25 percent, depending on rank and location.[44] One lieutenant general in Beijing reportedly received a raise of RMB 400 ($48) per month, while two senior colonels claimed 20 percent increases, from RMB 1700 to RMB 2040 ($205 to $290).[45] Overall, the average soldier in the PLA reportedly expected to receive an additional RMB 100 ($12) per month.[46]

Not surprisingly, the divestiture process still encountered resistance among military units reluctant to part with their enterprises. Some departments reportedly attempted to hide their enterprises under subordinate institutions that were not being screened by the central authorities.[47] Others tried to shield their profitable enterprises while willingly sacrificing their bankrupt ones. In cases where an enterprise was using the label "military enterprise" (jundui qiye) as a convenient cover for tax reductions and privileged access to transport or raw materials, individuals or units tried to have it reclassified as a nonmilitary enterprise. A significant number of enterprises were reportedly transferred to the control of relatives of military officers or defrocked military officers, thus retaining their unofficial links to their former units. Some of this backsliding was considered so serious that the office of the leading small military group in charge of the divestiture process was

---

[44] "Military Reportedly Raises Pay to Avoid Discontent," *Ming pao,* January 23, 1999, p. 15, in FBIS-CHI-99–023, January 23, 1999.

[45] Interviews in Beijing, February 1999.

[46] Kuang Tung-chou, "Beijing To Comprehensively Raise Servicemen's Remuneration," *Sing tao jih pao,* November 25, 1998, p. B14, in FBIS-CHI-98–349, December 15, 1998.

[47] Willy Wo-Lap Lam, "PLA Cashes In Its Assets," *South China Morning Post,* July 29, 1998.

forced to dispatch four work groups of 30 members each to inspect the larger units in the first half of December 1998.

As the divestiture entered 1999, additional serious bureaucratic and political conflicts began to surface. Overall, these can be divided into two categories: resource allocation and discipline. In terms of resource allocation, the PLA's expected boost in the 1999 official budget was far less than the military expected. In March 1999, Minister of Finance Xiang Huaicheng announced the military budget for the new fiscal year in his annual work report:

> In line with the CPC Central Committee request, central finances will provide appropriate subsidies to the army, armed police force, and political and law organs after their severance of ties with enterprises. In this connection, 104.65 billion renminbi, up 12.7 percent from the previous year because of the provision of subsidies to the army and of regular increases.[48]

Outside observers immediately noticed the meagerness of this figure, both in relative and absolute terms. At a relative level, the 12.7 percent increase was not significantly higher than the 12 percent increase of the previous year, calling into question the notion that the fiscal priority of the PLA had been raised. Even in absolute terms, the increase of RMB 13.65 billion ($1.6 billion) between 1998 and 1999 was not that much larger than the RMB 10.43 billion ($1.3 billion) increase between 1997 and 1998, and it reportedly included compensation of only RMB 3 billion ($362 million) for the loss of business income.

Initial interviews, bolstered by a loud chorus of PLA grumbling and complaining in official media, suggested that the military had been "duped" by the civilian leadership, which had implicitly promised a higher level of compensation.[49] The actual figure of RMB 3 billion in

---

[48] PRC Finance Minister Xiang Huaicheng, "Report on the Execution of the Central and Local Budgets for 1998 and on the Draft Central and Local Budgets for 1999," *Xinhua Domestic Service,* March 18, 1999, in FBIS-CHI-1999–0320, March 18, 1999.

[49] Interviews in Beijing with knowledgeable Chinese and Western interlocutors, 1999–2001.

compensation was instead based on the conservative profit estimate of RMB 3.5 billion ($0.42 billion on total revenue of RMB 150 billion [$18.1 billion]) that the PLA gave to Zhu Rongji before the divestiture announcement in July. This low estimate was very much in line with previous PLA estimates by the General Logistics Department, which consistently undervalued the profits from the military enterprise system to lessen the central tax burden of the commercial units. Thus, it appeared that the PLA was hoisted by its own petard. Ironically, interviews suggest that a genuine civil-military split over inadequate levels of budget compensation was averted thanks to the accidental bombing of the Chinese embassy in Yugoslavia in May 1999, which happened to occur during the programming phase of the Ninth Five Year Plan and resulted in a much more generous long-term budget trajectory than had previously been planned.[50]

Apart from budgets, additional resource allocation disputes arose over the distribution of enterprise assets in the post-divestiture environment. According to one well-informed observer, there were serious differences over levels of asset compensation because of the escalating costs of debts and liabilities incurred by enterprises.[51] Many firms were poorly managed with incomplete accounting records and had borrowed from multiple creditors. The firms' relationships with banks needed to be clarified, and business licenses needed to be reregistered. Another problem involved personnel. When the former military officers and other workers employed in these companies were transferred to the localities, their healthcare and insurance costs were transferred as well, creating unwanted social welfare burdens for the localities, i.e., the new owners.

The second major set of problems resulting from divestiture involved discipline issues, mainly corruption and profiteering. There is

---

[50] Interviews in Beijing with knowledgeable Chinese and Western interlocutors, 1998–2001. Some rough, back-of-the-envelope calculations suggest that the PLA was programmed for 20 percent annual increases over the five-year period, adjusted the following March for inflation. This represents a significant increase from the Eighth Five Year Plan, which was programmed at 10 percent.

[51] Personal communication with Tai Ming Cheung, September 9, 1999.

some evidence to suggest that the civilian leadership aggressively pursued investigations involving corruption in PLA enterprises, much to the chagrin of PLA officers who felt that the effort was gratuitous and harmful to the public reputation of the military.[52] Susan Lawrence and Bruce Gilley of the *Far Eastern Economic Review* reported from well-placed Chinese sources that the State Economic and Trade Commission (SETC) Receiving Office maintained a list of 23 company executives at the rank of major-general or above who had fled the country since 1998.[53] Seven of these officers were from the Guangzhou Military Region, which handed over more than 300 enterprises, and another five were from PLA headquarters units in Beijing. Among the latter is Lu Bin, former head of the General Political Department's Tiancheng Group, who was arrested overseas and extradited in January 2003. Other arrestees include a senior colonel who was the head of one of the PLA's top hotels, the Huatian, which is located in Changsha.

On December 15, 1998, the government officially announced the end of the second phase of divestiture, involving the formal registration and assessment of assets of the enterprises, followed by the expected official transfer of these enterprises to handover offices at the state, provincial, autonomous district, and municipality level.[54] Reportedly, 2,937 firms belonging to the PLA and People's Armed Police were transferred to local governments, and 3,928 enterprises were closed.[55] The big loser was the GLD, which saw more than 82 percent of its enterprises transferred or closed. One-third of the companies and their subsidiaries were retained by divestiture offices at the central level; the remaining two-thirds were transferred to divestiture offices at the local level. Profitable regional military conglomerates, such as the Jinling Pharmaceuticals Group in the Nanjing Military Region (MR), were

---

[52] Lawrence and Gilley, 1999, p. 24.

[53] Lawrence and Gilley, 1999.

[54] Cao Haili, op. cit., p. 5.

[55] "March Out of Business Sea: PLA and Armed Police Carrying Out the Decision of Divestiture," *Shidai chao,* March 2000.

placed directly under the direction of the regional commission.[56] By contrast, the ten mid-sized firms and 40 small firms owned by the Strategic Rocket Forces, whose businesses had not been terribly profitable, were given to the local governments.[57]

The commercial elements of China's most profitable military conglomerates, such as Xinxing, Songliao, and Sanjiu (999), were not handed over to local governments for reorganization but were instead placed directly under the control of the State Economic and Trade Commission in Beijing.[58] Eventually, it was thought that these large companies would become independent, state-owned conglomerates. The experience of Xinxing in this process is representative of the fate of these big firms.[59] Because Xinxing contained enterprises engaged in both military and nonmilitary production, its handover was very complicated. In the end, 56 numbered factories, which produce machines, logistics materials, clothes, and hats for the PLA, were kept under military control, but the trade group was transferred to the SETC. The ten specialized firms owned by Xinxing were reduced to seven after divestiture. Xinxing Foundry was retained by the military, and two other firms were transferred to conglomerates in the chemical industry. At the same time, three new firms, including the General Logistics Construction Company, which built the Military Museum, the new CMC Building, and the Beijing 301 Hospital, were added to Xinxing, restoring the number of firms to ten.

All large firms were subject to a broad set of rules. The central government would still control the nomination of the leadership of large firms and industrial groups and major enterprises of important

---

[56] Jin Weixin, "The Jiangsu Forces in the Nanjing Military Region Turn Over All Army-Run Enterprises, Stressing Politics, Considering the Overall Order, and Acting in Line With High Standards and Strict Requirements," *Nanjing Xinhua Ribao,* February 10, 1999, pp. 1,3, in FBIS-CHI-1999–0222, February 10, 1999.

[57] Cao Haili, op. cit., p. 9.

[58] First rumors of Songliao's transfer began to appear in late July. See Christine Chan and Foo Choy Peng, "Jiang Demand Threatens PLA Business Empire," *South China Morning Post,* July 24, 1998.

[59] This account is taken from Cao Haili, op. cit., p. 9.

industries. In terms of accounting and budget, the Ministry of Finance would manage the financial affairs of those firms under the control of the central government. All firms were required to participate in local social insurance schemes according to geographic divisions.

The remaining 8,000–10,000 PLA enterprises, most of which were the smaller, subsistence-oriented enterprises at the local unit level, remained under the ownership of the PLA.[60] The reforms were also "suspended" in some sectors, especially civil aviation, railway and posts, and telecommunications, because of the "special nature" of these industries.[61] For example, the Air Force's China United Airlines was permitted to continue operating.[62] Other notable exceptions included the 56 numbered factories previously under the control of the GLD's Xinxing Group, which remained under the administrative control of the GLD's pared-down Factory Management Department (formerly the larger Production Management Department), and Poly Group, which was divided between the General Equipment Department (arms-trading elements like Poly-Technologies), and the State Commission on Science, Technology, and Industry for National Defense (COSTIND).

In a sense, therefore, divestiture brought the PLA full circle. The pattern of the campaign, ranging from the transfers of its high-profile commercial enterprises to the retention of its lower-level farms and industrial units, suggests that the PLA's involvement in the economy has essentially returned to its pre-1978 "self-sustaining" pattern. Thus, the widespread conclusion that the PLA has been "banned" from business is far too simplistic. The military will continue to operate a wide variety of small-scale enterprises and agricultural units with the goal of supplementing the incomes of active-duty personnel and their dependents at the unit level. Profit and international trade, however, will no longer be critical features of the system. Moreover, the military leadership hopes

---

[60] Personal communication with Tai Ming Cheung, 24 January 1999.

[61] "Separation of Army from Business Done," *China Daily,* 21 March 1999, p. 1.

[62] China United Airlines reportedly survived divestiture because many remote towns protested that the shutdown on the airline would cut them off from the rest of the country.

that the divestiture of profitable companies will greatly reduce the incidence of corruption and profiteering in the ranks and thereby refocus the PLA on its important task of making itself more professional.

It is too soon to judge the long-term impact of this divestiture on the PLA. While participation in business had spawned endemic levels of corruption, an honest assessment would also admit that the military-business complex subsidized an underfunded military, improved the material life of the rank and file, and created jobs for cadre relatives. Despite these benefits, however, the military and civilian leaderships decided that the disadvantages of commercialism outweighed the advantages, particularly with the prospect of professional tasks like the liberation of Taiwan and potential military conflict with the United States on the horizon.

What will the short- to medium-term future hold for the divestiture process? Most likely, the next few years will witness repeated mop-up campaigns on the part of the central leadership and significant resistance and foot-dragging on the part of local military officials, repeating the pattern of earlier rectifications. An audit in early 1999 revealed that the military had kept some 15 percent of its businesses, necessitating the extension of some deadlines until August 1999. As late as May 2000, a top-level meeting on divestiture all but admitted that the military continues to shield some assets from the process, stating that the withdrawal of the military from business activities had only been "*basically* completed" (emphasis added).[63] Nonetheless, it is critical not to downplay the importance of what has already occurred. There is significant evidence to suggest that divestiture has ended the legal participation of the PLA in commercial activity, closing perhaps one of the most unique and interesting chapters of the post-Mao economic revolution.

---

[63] Wang Yantian and Yin Hongzhu, "CPC Central Committee, State Council, Central Military Commission Hold TV, Telephone Meeting on Work of Withdrawing Military, Armed Police, and Political and Legal Organs from Business Activities," *Xinhua Domestic Service*, May 25, 2000.

Estimating Actual Chinese Military Expenditure: A Bridge Too Far?

The Empirical Challenge

Given the paucity of verifiable data, estimates of the PLA's "total" budget have varied by orders of magnitude. Estimates of the 1994 budget, for example, range up to 24 times the official PLA outlay of $6.3 billion, as seen in Figure 4.10.

Data Challenges and the PPP Debate

Above and beyond the issue of paucity, using the existing official Chinese data can be problematic. As Gill points out, official sources "often employ inconsistent terms and definitions to explain defense spend-

Figure 4.10
Estimates of Actual 1994 Military Expenditures

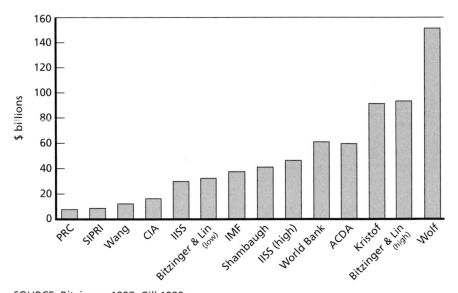

SOURCE: Bitzinger, 1997; Gill 1999.
NOTES: ACDA = U.S. Arms Control and Disarmament Agency.
RAND MG260-4.10E

ing."[64] Moreover, converting expenditures in renminbi into dollars poses difficult problems. Clearly, a given dollar has different levels of relative spending power depending on what is being purchased. As Richard Bitzinger points out, "many goods in the Chinese 'defense spending basket' cost much less in China than in the West: low pay and poorer living standards in the PLA saves monies on personnel, and lower wages at defense factories depresses the costs of arms procurement."[65] As a result, he argues that "these disparities should be corrected by some kind of PPP multiplier, especially when attempting to compare Chinese defense spending to military expenditures in other countries."[66] Below, we use a purchasing power parity exchange rate to convert military expenditures on personnel into 2001 dollars. We argue that the PPP exchange rate provides a more accurate reflection of the true value of these military expenditures in terms of relative spending power.

However, as noted in Chapter Two, PPP exchange rates are problematic for a number of reasons, especially in the case of military spending. PPP calculations often require unobtainable data about military prices. They may fail to account for the inferior quality of Chinese products or services relative to those in the West. In the case of imports of military equipment, market exchange rates are a better measure of dollar value because these items are purchased at world market prices. For all these reasons, PPP-based analysis must be used carefully and appropriately caveated, because it may overvalue Chinese expenditures. In Chapter Seven, we make judicious use of PPP exchange rates in our forecasts of future military spending levels.

How Do We Fill In the Gaps?

Newly acquired internal materials offer a relatively complete picture of the budgeting and auditing calendar. Four categories of data appear to be "low-hanging fruit." The first is foreign arms purchases, many of which are overt and calculable. The second is financial support to local

---

[64] Gill, 1999, p. 195.

[65] Bitzinger, 2003, pp. 164–175.

[66] Bitzinger, 2003.

units from civilian government institutions. Newspapers and National People's Congress (NPC) discussions often air complaints by government officials about the burden of supporting soldiers and reserves. Provincial statistical yearbooks also offer a wealth of data. The other obvious category for examination is paramilitary expenditures for the People's Armed Police, which can be found in Ministry of Civil Affairs statistical yearbooks, as well as Ministry of Public Security newspapers and journals. In examining these civilian data sources, Wang Shaoguang offers a useful angle of attack, highlighting that past efforts to example specific military budget categories did not take the simple step of cross-checking expenses with known civilian data. As a result, for example, some published estimates of military pensions exceeded the entire state pension allocation in the national statistical yearbook.[67] Finally, detailed data is available on China's arms sales to other countries, which reportedly generates some commissions for the PLA. Each of these will be dealt with in turn below.

Foreign Arms Purchases

In the past ten years, the Chinese have purchased a significant amount of military equipment from foreign suppliers, particularly Russia. The main systems acquired include Su-27 Flanker and Su-30MKK fighter aircraft, Sovremenny-class destroyers, Kilo-class diesel submarines, and SA-10/15/20 surface-to-air missiles. These purchases are commercial transactions and are negotiated on commercial terms. One U.S. government estimate placed the level of spending on foreign arms at $800 million per year in the early 1990s. Later estimates concluded that expenditure had risen to $1.5 billion per year in the late 1990s, growing to a level of $3.6 billion per year in 2002.[68]

Provincial Spending on National Defense

From provincial statistics, it is possible to analyze the level of provincial contribution to national defense (*guofang*). Figure 4.11 reveals the

---

[67] Wang, 1996, pp. 889–911.

[68] The latest estimate from Robert Grimmett at Congressional Research Service is quoted in Jim Wolf, "China Stays Top Weapons Importer: U.S. Report," Reuters, 27 September 2003.

Figure 4.11
Provincial Contributions to National Defense ($ millions)

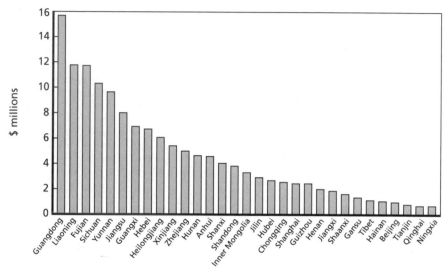

SOURCE: *CSY*, 2003.
RAND *MG260-4.11*

support levels for 2001. Total provincial spending on defense affairs in 2001 was $800 million, well below previous Western estimates. Indeed, the outlays in individual provinces make up a surprisingly small percentage of local budgets, ranging from 0.015 percent to 0.244 percent, as seen in Figure 4.12.

The biggest-spending province was Fujian, one of the most important "war zones" for PLA operations. As a result, the localities in this area are likely burdened with disproportionately high expenditures to cover the costs of hosting major exercises. At the other end of the spectrum, the local government outlays for Beijing, Shanghai, and Tianjin are strangely low, given the government's keen interest in maintaining internal stability, but perhaps reflect funding of military units from the central government or the headquarters units.

Figure 4.12
Provincial Contribution to National Defense as a Percentage of Total
Provincial Government Spending

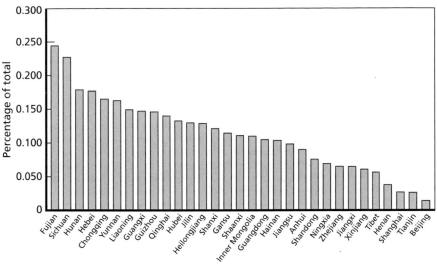

SOURCE: *CSY*, 2002.
RAND *MG260-4.12E*

Paramilitaries

There are three main sources of funding for the paramilitary forces, the People's Armed Police. The first is "state finances," including money from the central civilian budget and the budgets of central departments or ministries that maintain specialized PAP units. PAP has a distinct budget line in the Minister of Finance's Budgetary Report and Premier's Government Work Report. The ministries or departments providing support include public security, justice, forestry, transportation, and the departments of the industrial ministry responsible for nonferrous metals such as gold.

The second major source of funding is known as "local finances" and includes money for units located at the same level of provincial, municipal/prefectures, or county governments. Provincial expenditures for PAP units in 2001 totaled $213 million.

The money spent on paramilitary units as a percentage of provincial budgets is again relatively low, ranging from 0.03 percent to 0.36 percent of total provincial expenditures (Figure 4.13). The absence of Beijing

Figure 4.13
Provincial Contributions to Paramilitary Forces as a Percentage of Total
Provincial Government Spending

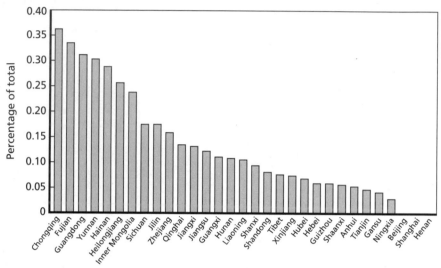

SOURCE: *CSY,* 2002.
RAND *MG260-4.13E*

and Shanghai data is surprising, particularly given the high priority for internal security in the capital. The top provinces include coastal areas that require PAP coast guard units, border police, and customs officials, such as Guangdong, Fujian, Zhejiang, and Jiangsu, as well as border areas requiring a significant nonmilitary presence. Yunnan and Guangxi, for example, are top smuggling corridors, and the PAP is on the front lines of the drug war opposite the Golden Triangle in Southeast Asia. Heilongjiang and Inner Mongolia require extensive border policing, and Jilin and Tibet are marked by worker unrest and restive populations.

The final category is extra-budgetary funds from PAP businesses, fines, and security fees from government units and enterprises. The latter category largely includes fees paid by ministries in Beijing for PAP protection of their facilities and housing. There is no evidence that the PAP receives any funding from the official PLA budget, although the paramilitaries are reportedly under the operational control of the General Staff Department during crises and would likely have their operational expenses paid by military sources. Since the PAP remains organizationally under the State Council, civilian central units have responsibility for salaries and benefits.

Revenue from Arms Sales

China is a relatively minor player in the global arms market. Figure 4.14 shows that revenue in the 1990s ranged from $400 million to $1200 million per year.[69]

The main customers have been developing nations, including Pakistan, Burma, and Thailand. The revenue from these sales goes primarily to defense-industrial firms, although the PLA reportedly receives a small commission for the transaction and additional monies if the sale involves surplus small arms from PLA warehouse stocks.

Figure 4.14
Chinese Revenue from Arms Sales

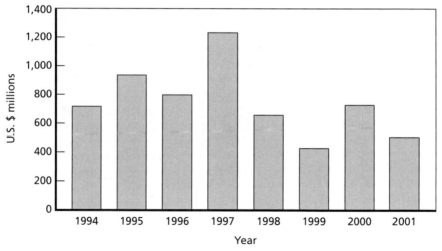

SOURCE: Grimmett, 2002.
RAND *MG260-4.14*

---

[69] For the sake of contrast, the value of arms and military equipment sold by the United States in 2001 was 19 times as much as that sold by China in terms of monetary value.

Bounding Other Categories of Indirect Funding

Two final categories of indirect defense revenue are state subsidies to defense-industrial enterprises and monies for defense research and development. In the absence of identifiable disaggregated data in these areas, it is reasonable to establish an upper bound by examining overall levels of subsidies to state-owned enterprises and state spending on research and development, since Chinese sources explicitly define these categories as including defense components. Outlays on subsidies to loss-making SOEs were $3.1 billion in 2002. Funds for R&D were $4.3 billion in 2001, as shown in Figure 4.15.[70] As documented in detail in Chapter Five, the upper-bound figure for direct subsidies substantially exceeds actual payments to these industries. For example, in 2000, the defense sector reportedly received roughly $800 million in direct subsidies, about a quarter of the total.[71] Our lower-bound esti-

Figure 4.15
State Subsidies to State-Owned Enterprises and State Funding of R&D

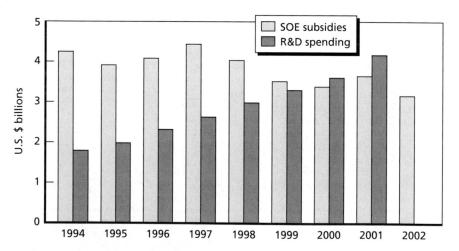

SOURCE: *CSY*, 2002, 2003, Tables 8.3 and 8.10.
RAND *MG260-4.15E*

---

[70] *China Statistical Yearbook 2002*, pp. 278, 270.

[71] Fu Jing, "Defense Industry Eyes Foreign Cash," *China Daily*, 4 July 2001, in FBIS as "Chinese Defense Industry Eyes More Foreign Investment," 4 July 2001.

mate of direct subsidies to defense enterprises for 2001 is based on this number. If anything, direct budgetary subsidies to the defense industry appear to have declined since 2000. In 2002, the defense industry reportedly broke even because, on a net basis, earnings from profitable operations covered losses in other industries.[72] Prior to 2001, the "aerospace" (i.e., missile) sector was the only profitable segment of China's defense industry.[73] Both of these numbers for total state spending are lower than previous outside estimates of the defense subtotals.

A Notional Full Budget

From the data presented above, it is possible to make a first cut at creating an upper bound on Chinese military expenditures (Table 4.1).

The base is the official defense budget of $22.4 billion for 2003. On top of this one can add estimates of Chinese imports of military equipment ($3.6 billion), provincial support to national defense ($1.18 billion), and paramilitary expenses ($3 billion). Finally, we assume the defense-industrial subsidies and R&D funding are bounded at $0.8

Table 4.1
Notional Full Chinese Military Budget

Category	Amount ($ billion)
Official budget	$22.4
Funds for foreign arms imports	$3.6 (2002)
Local support to defense and paramilitaries	$1.13 ($800 million for defense, $213 million for paramilitaries)
Defense industrial subsidies	$0.8 to $3.1
Defense R&D	< $4.3
Paramilitaries	$3
Arms sales revenues	< $0.5
Total funds	~$31–$38

[72] "Defense Industry Breaks Even in 2002," *China Daily* (internet version), 9 January 2002.

[73] Zhang Yi, "It Is Estimated that China's Military Industrial Enterprises Covered by the Budget Reduce Losses by More Than 30 Percent," Xinhua Hong Kong Service, 7 January 2002, in FBIS as "PRC Military Industry Reports 30 Percent Decrease in Losses for 2001," 7 January 2002.

billion to $3.1 billion and $4.3 billion, respectively, and that the PLA portion of arms sales revenue must be less than the defense industry's total revenue. Using this combination of data and assumptions, the total defense expenditures must run between $31–38 billion per year, or 1.4 to 1.7 times the official number.

Extrapolating this difference backwards, Figure 4.16 shows notional PRC defense spending from the period 1987 to 2003.

While this comparison is useful perhaps only for heuristic purposes, it does highlight possible important trend lines, especially when compared with the recent noticeably downward tilt in Taiwan defense spending over the same period.

Figure 4.16
Revised Notional Chinese Defense Spending, 1987–2003

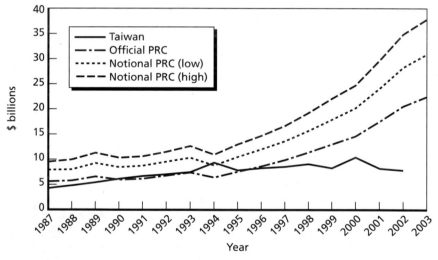

CHAPTER FIVE

China's Defense Industry

Capabilities and Future Prospects

Up to this point, we have focused on the resource side of the military budget equation. We have evaluated past and likely future trends in the Chinese economy and government budgets, explored the military budgeting process in detail, and provided alternative estimates of Chinese military spending, incorporating off-budget categories. However, we have yet to evaluate what the Chinese government is able to purchase with these resources. In this chapter, we assess the strengths and weaknesses of China's defense industries and evaluate likely future prospects. The chapter is designed to address the question of whether the Chinese government will be able to translate increased expenditures into purchases of reliable, technologically sophisticated weaponry.

Our conclusions are more positive than those of most Western analyses to date. Over the past 20 years, one of the most prominent and consistent conclusions drawn from research on China's defense industrial complex has been that it is rife with weaknesses and limitations.[1] We find that China's defense industries are producing a wide range of increasingly advanced weapons that, in the short-term, would enhance not only China's capabilities in a possible conflict over Taiwan but also China's long-term military position in Asia. From the vantage point of 2003, we argue that it is time to shift the focus of research to the gradual improvements in and the future potential of China's defense industrial complex rather than concentrating solely on its weaknesses.

---

[1] Gill, 2001.

As noted above, the Chinese government has been substantially increasing its reported military spending, especially on defense procurement. Reported expenditures on procurement increased by an average of 21 percent per year between 2000 and 2002.[2] These reported increases in expenditures have been matched by purchases of more and better weaponry, an appreciable share of which is manufactured by Chinese enterprises. These weapons and equipment reflect improvements in the technological capabilities of China's defense manufacturing base. Part of these improvements stem from a stronger resource base. China has a growing pool of technical talent in its civilian sector whom Beijing is now attempting to attract to work in the defense sector. The government is also making a concerted effort to reform the institutional framework and incentives under which the defense industry operates. These reforms are taking time, but a number of initial indications suggest that progress is occurring, especially compared with the previous rounds of rather feckless attempts at defense industry reform that have occurred since 1978. In the words of General Li Jinai, the head of the General Armaments Department (GAD), the lead defense procurement agency, "there has been a marked improvement in national defense scientific research and in building of weapons and equipment. The past five years has been the best period of development in the country's history."[3] These trends bear close watching by U.S. policymakers, analysts, and military planners.

As China's economic and resource base expands in the coming years, Beijing has three broad paths by which to translate these economic achievements into improved military capabilities. The first is to produce all the weapons needed to equip the country's military domestically. The second is to purchase major weapons from the high-tech military equipment manufacturers of the world, such as the United States, Russia, Britain, France, Germany, Sweden, and Israel. A third

---

[2] See FBIS, "*Xinhua:* 'Full Text' of White Paper on China's National Defense in 2002," December 9, 2002. This figure is believed to exclude expenditures on military equipment imported from Russia or other countries and expenditures on research and development activities.

[3] Wang Wenjie, "Delegate Li Jinai Emphasizes: Grasp Tightly the Important Strategic Opportunity, Accelerate the Development By Leaps of Our Army's Weapons and Equipment," *Jiefangjun Bao,* March 8, 2003, p. 1, as translated in FBIS, March 8, 2003.

path combines these two approaches by attempting to improve domestic manufacturing processes and military equipment designs to produce better quality weapons at home while importing from abroad weaponry and equipment that domestic manufacturers are not yet capable of producing. After largely pursuing the first path for much of the 1960s, 1970s, and part of the 1980s, since the 1990s China has been following the third path—improving domestic production while purchasing growing numbers of advanced weapon systems from abroad, mostly from Russia and Israel.

China's leaders and strategists do not like being dependent on other countries for their defense modernization needs. They have made it clear that their long-term goal is to return to the first path—"self-reliance" in defense production. As the volume of Russian imports in recent years indicates, however, China's military has been decidedly unsatisfied with the quality of the products of China's defense industries at a time when the PLA is accelerating its efforts to develop real military options in the event of a conflict over Taiwan. The ability of the PLA to overcome its self-proclaimed problem of "short arms and slow legs,"[4] depends on the ability of China's defense R&D institutes and factories to overcome their past inadequacies and produce sophisticated and reliable weapons systems.

This chapter begins by reviewing the evolution of China's defense industrial policies since the beginning of Chinese economic reform efforts in the late 1970s. China's most recent round of defense industry reforms initiated in the late 1990s is given particular attention. The chapter then examines the organization and production capabilities of four key sectors of China's defense industry: aviation, shipbuilding, information technology and defense electronics, and missiles. The chapter concludes with an evaluation of the future production capabilities of China's defense industries over the next two decades and their potential contributions to military modernization.

Both this overview section and each of the case studies focus heavily on changes in two key factors: the structure of *institutions* and the

---

[4] Since the mid-1990s, China's PLA Daily *(Jiefangjun Bao)* has been using this phrase to describe the military's limitations.

nature of *incentives* for increased efficiency and innovation. Changes in both these factors in the defense industry have begun to gradually re-shape the operations and output of China's defense industries. Changes in these two factors are examined at the levels of *government* operations (i.e., defense procurement decisions) and *enterprise* operations. By focusing on institutions and incentives at both the government and enterprise level, this chapter provides a means by which to analyze and evaluate potential progress of China's defense industrial establishment in improving the quality and capabilities of future weapons systems.

The Changing Shape of China's Defense Industrial Complex, 1980–1998

From the late 1970s, when Deng Xiaoping initiated reform of China's planned economy, until recently, China's defense industries[5] led a troubled existence. Government procurement of military goods declined dramatically following the adoption of Deng's "Four Modernizations Policy" which placed the military as the last priority.[6] As a result, many defense enterprises were officially encouraged to convert their facilities to the production of nonmilitary goods or engage in arms sales to generate income to replace dwindling government purchases of military equipment. Many firms soon became dependent on these alternate sources of income for their very survival.

Defense conversion (军转民) was a troubled process for most Chinese firms. Despite Chinese government claims to the contrary, China's weapons producers found it difficult to shift production to goods that could be profitably sold on the booming domestic market. Military good producers were hampered by legal constraints and difficulties in attracting foreign partners who could provide new capital, know-how,

---

[5] China's "defense enterprises" (军工企业), state-owned industries that produce weapons and equipment for China's military, are distinct from "military enterprises" (军队企业), which used to be owned and operated by PLA until divestiture by the PLA in the late 1990s. See Mulvenon, 1997a. In July 1999, Jiang Zemin called for the PLA to sever its ties to all business enterprises. This process is ongoing.

[6] See Frankenstein, 1999.

and technologies. They also lacked the managerial flexibility to replicate the successes of the many new Chinese companies that emerged during the reform period. These problems were exacerbated by the general weaknesses of China's state-owned enterprises in terms of their ability to absorb new technologies, improve management, and improve the technical skills of their workforce. The Chinese government's commitment to self-reliance in military equipment production also hindered the ability of these enterprises to successfully sell to nondefense markets because factories had to remain capable of producing a full range of components and equipment for military production, thus preventing specialization and the accompanying increases in quality and technological sophistication that longer production runs potentially provide. As a result, many civilian goods produced by defense firms have been of low quality and relatively costly to produce and thus failed to generate profits.[7]

These problems are becoming more acute because the entry of many new Chinese companies in all markets has intensified domestic competition. Competition from foreign producers has also emerged now that China has entered the WTO and China's domestic markets for consumer goods have become more open to foreign manufacturers. Although some sectors, such as shipbuilding and electronics, demonstrated an impressive ability to transform themselves into profitable firms producing mainly nonmilitary goods, few defense firms in other sectors were able to use defense conversion to gain access to modern production technologies to upgrade their facilities for producing better military goods—the gov-

---

[7] Zhang Yihao and Zhou Zongkui, "China's Science, Technology, and Industry for National Defense Face up to WTO—An Interview With Liu Jibin, Minister in Charge of the Commission of Science, Technology, and Industry for National Defense," *Jiefangjun Bao* (internet version) 13 March 2000, p. 8, in FBIS as "WTO Impact on PRC Defense Industry," 14 March 2000; Wang Jianhua, "Thoughts on 'WTO Entry' and Development of Armament," *Jiefangjun Bao* (internet version), 14 March 2000, p. 6, in FBIS as "Impact of WTO on PRC Armament Development," 15 March 2000; Ke Wen, "Advantages and Disadvantages of WTO Accession to China's Military Industry, Science and Technology—Interviewing Liu Jibin, Minister in Charge of Commission of Science, Technology, and Industry for National Defense," *Chiao Ching*, 16 June 2000, pp. 46–48, in FBIS as "Minister Liu Jibin on Pros, Cons of WTO Accession to PRC Defense Industry," 20 June 2000. See also Frankenstein, "China's Defense Industries: A New Course?" op. cit.

ernment's initial rationale for encouraging defense conversion.[8] At best, defense conversion has had mixed success in China.[9]

Weapons sales did become a substantial source of income for some defense firms in the 1980s. Chinese arms sales peaked in 1987 at over $1 billion. This source of revenue quickly dried up in the early 1990s, however, after demand disappeared following the cessation of the Iran-Iraq war. Chinese weapons exporters also lost export markets because of the very poor performance of their weapons in Iraq's hands during the 1991 Gulf War and because of competition from the influx of technologically superior and relatively inexpensive Russian weapons onto international markets following the collapse of the Soviet Union.[10] China's expanding arms control and nonproliferation commitments gradually curbed exports of its most competitive and desired military systems such as surface-to-surface and antiship missiles.[11] As a result of the worsening environment for weapons exports, China's defense firms, like many of China's large state-owned enterprises (*guoyou qiye* 国有企业), staggered along for most of the 1990s by relying on significant government subsidies.[12] As late as 2000, China's defense industries were described as facing "tremendous challenges."[13] The difficult circumstances faced by China's

---

[8] Liu Jibin, "Implement the Guideline of Military-Civilian Integration, Rejuvenate the National Defense Science and Technology Industry," *Renmin Ribao,* 2 February 1999, p. 12, in FBIS as "Military-Civilian Integration in Industry," 2 February 1999. See also Frankenstein, 1999; Brommelhorster and Frankenstein (eds.), 1997; Frankenstein and Gill, 1996, p. 394–427; and Gill, 1996, pp. 144–167.

[9] Medeiros, 1998.

[10] Eikenberry, 1995; Gill, 1992; Bitzinger, 1992.

[11] Medeiros and Gill, 2000.

[12] Xinhua, 7 January 1998, in FBIS as "PRC National Ordnance Industry Conference Opens," January 7, 1998; Bie Yixun, Xu Dianlong, Xinhua Domestic Service, 7 January 1998, in FBIS as "Wu Bangguo Greets Opening of Ordnance Industry Meeting," 12 January 1998. In 2000, the defense sector reportedly received RMB 1.7 billion in direct subsidies, in addition to an uncertain amount of indirect subsidies in the form of government-directed loans from state-owned banks. Fu Jing, "Defense Industry Eyes Foreign Cash," *China Daily,* 4 July 2001, in FBIS as "Chinese Defense Industry Eyes More Foreign Investment," 4 July 2001.

[13] Ke Wen, "Advantages and Disadvantages of WTO Accession to China's Military Industry, Science and Technology—Interviewing Liu Jibin, Minister in Charge of Commission of Science, Technology, and Industry for National Defense, " op. cit.

defense industrial enterprises were reflected by their financial circumstances. Despite the fact that supposedly over 80 percent of the aggregate output of Chinese defense enterprises was civilian goods,[14] few firms were "profitable." According to the director of the COSTIND, China's entire defense industry ran a net loss in aggregate terms for eight consecutive years—from 1993 to 2001.[15]

Aside from the financial problems of China's defense industries, the military systems they have produced have been unimpressive. The weaknesses of China's defense production capabilities are reflected by the technological backwardness of many of the systems, the long R&D and production timelines for most indigenously built weapons, and China's growing reliance on purchases of major weapons systems from foreign countries. The history of China's defense industry is replete

---

[14] Liu Jibin, "Implement the Guideline of Military-Civilian Integration, Rejuvenate the National Defense Science and Technology Industry," op. cit., Tang Hua, "Science, Technology, and Industry for National Defense Increases Intensity of Innovation," *Liaowang,* 26 July 1999, pp. 16, 17, in FBIS as "Report on Innovation in Defense Industry," 16 August 1999; Zhang Yihao, Zhou Zongkui, "China's Science, Technology, and Industry for National Defense Face up to WTO—an Interview With Liu Jibin, Minister in Charge of the Commission of Science, Technology, and Industry for National Defense, op. cit.; Ke Wen, "Advantages and Disadvantages of WTO Accession to China's Military Industry, Science and Technology—Interviewing Liu Jibin, Minister in Charge of Commission of Science, Technology, and Industry for National Defense," op. cit.; Zhao Huanxin, *China Daily* (internet version), 19 December 2000, in FBIS as "China Daily: Military Technology Transfer To Spur Growth in Civilian Sector," 19 December 2000; Fan Rixuan, "The Profound Impact of China's WTO Accession on People's Lives and Thinking, As Well As on National Defense and Military Modernization Drive—Thoughts on China's WTO Admission and National Defense Building," *Jiefangjun Bao* (internet version), 30 April 2002, p. 6, in FBIS as "Article Discusses Impact of China's WTO Admission on National Defense Building," 2 May 2002.

[15] "Defense Industry Breaks Even in 2002," *China Daily* (internet version), 9 January 2002; Zhang Yi, "It Is Estimated that China's Military Industrial Enterprises Covered by the Budget Reduce Losses by More Than 30 Percent," Xinhua Hong Kong Service, 7 January 2002, in FBIS as "PRC Military Industry Reports 30 Percent Decrease in Losses for 2001," 7 January 2002. The "ordnance" (i.e., ground systems) sector reportedly "suffered serious losses" for over ten years beginning at the end of the 1980s—Jia Xiping, Xu Dianlong, "China's Ordnance Industry Achieves Marked Successes in Reform and Reorganization to Streamline and Improve Core Business," Xinhua Domestic Service, 19 January 2000, in FBIS as "PRC Ordnance Industry Reform Results," 10 February 2000; Qian Xiaohu, "Crossing Frontier Passes and Mountains With Golden Spears and Armored Horses—Interviewing Ma Zhigeng, President of China Ordnance Group Corporation," *Jiefangjun Bao* (internet version), 17 April 2000, p. 8, in FBIS as "China Ordnance Group Chief Interviewed," 17 April 2000. For additional information and details on the poor economic state of China's defense industry in the 1980s and 1990s see Frankenstein, 1999.

with examples of weapon systems with severe technological weaknesses and limitations. In particular, manufacturers appear to have difficulty in making technological leaps to qualitatively new, superior systems. While many new types of tanks, artillery, surface-to-air missiles, surface ships, submarines, and air-to-air missiles have entered service since 1980, for the most part these new designs have been incremental improvements on earlier ones, which in many cases can trace their lineage back to 1950s Soviet technology.

The limitations of China's defense industries are reflected in the long production cycles for major defense systems. China's JH-7 (FBC-1) fighter-bombers and J-10 (F-10) multirole aircraft, its most advanced indigenously produced military aircraft, were both under development for two decades. The JH-7 only recently entered into service for the PLA Navy (PLAN), even though it was first designed in the early 1970s. Despite the very long development times involved, the project is still dependent on jet engines imported from Britain—China has been unable to produce the engine on its own. The J-10 has just entered series production despite the fact that the program was initiated in the early 1980s, and the design is largely derived from Israel's canceled Lavi fighter program (which in turn was based on U.S. F-16 technology).[16]

Other sectors of China's defense industry have exhibited similar, though perhaps not as acute, weaknesses as the aircraft industry. For most of the 1980s and 1990s, China produced no heavy naval cruisers or multirole destroyers with advanced air defense or antisubmarine systems. Until recently, China's newest classes of surface ships were produced in very small numbers, showed few significant design innovations, and relied on imported equipment for critical subsystems, weapons, and sensor suites. Even China's missile sector, which is often heralded as a "pocket of excellence," does not inspire awe. The solid-fuel ballistic missiles and antiship cruise missiles for which it has made its reputation are comparable to systems fielded in the West in the 1960s and 1970s. The missile industry has experienced continued problems developing a long-range

---

[16] For a history of the troubled procurement cycles of these aviation programs see Allen, Krumel, and Pollack, 1995.

naval surface-to-air missile (SAM), and the resulting delays have complicated some of the planned upgrades of both current and future naval platforms. Furthermore, China's long-range ballistic missile modernization program, which began in the mid-1980s, has so far resulted in the production of only one new long-range, solid-fueled, road-mobile system (known as the DF-31) almost 20 years after its initial conception.[17] Most critically, China's defense industries have so far been unable to produce the types of radically new technologies, such as low observable systems and precision guided munitions, that the U.S. military has employed so effectively in recent years. Although the Chinese press is fond of boasting how Chinese weapons have reached "advanced world levels,"[18] numerous sources more soberly acknowledge that China's defense industry has persistently failed to meet the needs of the military and has lagged behind the technological development in the more market-oriented sectors of the Chinese economy.[19]

---

[17] For a useful review of the development cycles of major Chinese weapons systems see Shambaugh, 2003, especially Chapter Six, "Defense Industries and Procurement."

[18] See, for instance, Wang Fan and Zhang Jie, "A Qualitative Leap in the Overall Strength of National Defense over Past 50 Years," *Liaowang*, 26 July 1999, pp. 10, 11, 12, in FBIS as "Xinhua Journal Reviews PRC Defense Growth," 4 August 1999; Li Xuanqing, Fan Juwei, and Su Kuoshan, "Defense Science and Technology Forges Sharp Sword for National Defense—Second Roundup on Achievements of Army Building over Past 50 Years," *Jiefangjun Bao*, 7 September 1999, pp. 1, 2, in FBIS as "Overview of PLA Defense S&T Modernization," 21 September 1999; Yu Bin, Hao Dan "Qualitative Changes in National Defense Modernization Standard, Overall Strength," 27 September 1999, pp. 52–54, in FBIS as "Changes in National Defense, Power," 18 October 1999; Zhongguo Wen She, "China's Military Science and Technology Develop by Leaps and Bounds," *Renmin Ribao* (internet version), 27 July 2000, in FBIS as "PRC Claims Military Science, Technology Becoming More 'Dependable,'" 27 July 2000.

[19] Chen Zengjun, "Ordnance Industry Turns into Vital New National Economic Force," *Jingji Ribao*, 20 November 1998, p. 2, in FBIS as "Ordnance Industry Becomes 'Vital' Economic Force," 12 December 1998; Xinhua Domestic Service, 27 April 1999, in FBIS as "Wu Bangguo Speaks at Defense Industry Conference," 2 May 1999; Tang Hua, "Science, Technology, and Industry for National Defense Increases Intensity of Innovation," op. cit. An article in *Jiefangjun Bao (Liberation Army Daily)* characterized China's armaments as suffering from "one low" and "five insufficiencies": a low degree of informationization; insufficient high-power armaments; insufficient strike weapons; insufficient precision-guided weapons; insufficient means of reconnaissance, early warning, and command and control; and insufficient electronic weapons. An Weiping, "Thoughts on Developing Armaments by Leaps and Bounds," *Jiefangjun Bao*, 6 April 1999, p. 6, in FBIS, 6 April 1999.

Perhaps the strongest indictment of the failures of China's defense industries is the PLA's high and growing reliance on purchases of major weapons systems from foreign countries (mainly Russia). Since the mid-1990s, China has sought to fill critical gaps in its force structure not through indigenous procurement but rather by purchasing advanced weapons systems and related technologies from abroad. This foreign procurement has been in a broad range of categories of weapons systems. The most significant purchases have included advanced fourth-generation fighter aircraft, modern destroyers with advanced air defense and antisurface capabilities, long-range land-based air defense systems, advanced diesel-electric submarines, jet engines, and advanced defense electronics technologies. These numerous purchases of advanced foreign weapons systems serve as a strong indicator of the PLA's perceived capability gaps as a result of the inability of China's defense industrial base to meet the PLA's needs.

Explaining the Defense Industry's Poor Performance

The reasons for the slow technological progress in China's defense industries in the 1980s and 1990s are similar to the reasons for slow adaptation of new technologies in the rest of China's state-owned sector.[20] Chinese managers in both defense and civilian state-owned enterprises lack incentives to innovate. China's defense manufacturers were paid cost plus 5 percent for equipment they produced. Cost-plus reimbursement schemes provide no incentives for manufacturers to cut costs. Enterprise managers had little or no role in procuring contracts. Decisions about which company would produce a particular item were made by administrative fiat and ministerial bargaining rather than through competitive bidding among manufacturers.[21] As a result, military equipment producers had little financial interest in improving the quality of the weapons systems they produced or the efficiency with which they manufactured

---

[20] See Baark, 1991; Suttmeier, 1991, 1997; Zhou, 1995.

[21] Gill and Henley, 1996.

or designed them because improvements had little effect on the orders the company received or the profits it made.[22]

In addition to the lack of financial incentives for innovation, China's Soviet-inspired approach to industrial organization inhibited innovation. Under the Soviet model, R&D activities are performed by institutes that are organizationally separate from the actual manufacturers. This situation remained common, though not universal, in China's defense sectors during the 1980s and 1990s. The institutes were funded through annual budgetary allocations from the central government and received little direction from production enterprises in their research efforts. Not only did this result in technology development projects that were impractical or unrelated to production needs, it meant that research and development institutes were deprived of the valuable technological information generated in the course of working out the kinks in production processes at the plant level.[23]

Another drawback of the Soviet model is its top-down organization. Because directions for technological development were determined by the central authorities, technological opportunities that arise serendipitously but do not fit into these pre-determined directions tend to be neglected. The hierarchical organizational structure discourages the horizontal knowledge flows that are critical to technological progress.[24] This knowledge flow problem was undoubtedly exacerbated by the extreme secrecy associated with defense production in China.

Other problems of the defense industries included excessive production capacity, redundant personnel, inflexibility in hiring and firing, loss of quality personnel to the non-state-owned sector, incorrectly priced inputs, poor management practices, and the inefficient geographic distribution of industry resulting from a 1960s and 1970s

---

[22] In fact, reductions in the cost of production would actually *reduce* the profits that an enterprise received.

[23] Gallagher, 1987.

[24] Frieman, 1993, p. 54; Frankenstein and Gill, 1997, p. 80; Arnett, 1995, pp. 370–371; Ding, 1996b, p. 82; Baark, 1997, pp. 92, 106, 108.

policy of relocating defense industries to remote interior areas where they would be behind China's "Third Line" (大三线) of defense from an external invasion.[25]

Weaknesses of Past Reforms

The Chinese government has recognized the shortcomings of its defense industrial complex and, during the 1980s and 1990s, made repeated attempts to reform and rehabilitate it. These efforts relied mainly on two strategies: defense conversion, discussed above, and institutional reorganization. Similar to China's experience with defense conversion, institutional reorganization was largely a cosmetic and ineffective pathway to reforming China's defense production capabilities. The government frequently changed the names of enterprises and institutions and shuffled organizational responsibilities, but it took few of the institutional measures needed to consolidate and rationalize the industry so as to increase efficiency and foster innovation. A brief recitation of these efforts elucidates their weaknesses.

Before the early 1980s, China's defense industrial complex consisted of a series of numbered "machine building industries" (MBIs) representing the major defense sectors—such as nuclear weapons, aviation, electronics, "ordnance" (tanks, artillery, etc.), shipbuilding, and

---

[25] Bie Yixun, Xu Dianlong, Xinhua Domestic Service, op. cit.; Hsiao Cheng-chin, "Liu Jibin, Minister in Charge of the State Commission of Science, Technology, and Industry for National Defense and a Veteran Who Has Rejoined His Original Unit," *Hsin Pao*, 3 June 1998, in FBIS as "Article on New Minister Liu Jibin," 12 June 1998; Peng Kai-lei, "Five Major Military Industry Corporations Formally Reorganized," op, cit; Zhang Yi, "Liu Jibin, Minister of Commission of Science, Technology, and Industry for National Defense, 27 October Says in a Meeting that China's High-Tech Industry for National Defense Will Be Restructured on a Large Scale," Xinhua Domestic Service, 27 October 1999, in FBIS as "Official Urges Major Defense Industry Shakeup," 3 November 1999; Xinhua Domestic Service, "China's Military Industrial Industry Last Year Decreased Losses By a Large Margin," 5 January 2001, in FBIS as "PRC Says Military Industry Reduces Losses," 5 January 2001; Conroy, 1992; Frankenstein, 1993, 1999; Baark, 1997. On the Third Line (or "Third Front") policy the seminal work is Naughton, 1988. See also Conroy, 1992, pp. 63–64. One additional negative consequence of the Third Line policy on China's defense industries is the difficulty in attracting talented personnel to work in the isolated, backward regions where China's defense industries are often located. Ding, 1996b, p. 86.

"aerospace" (missiles and space). At that time, these industries were overseen by numerous organizations with overlapping responsibilities and claims to ownership including the State Planning Commission, the Ministry of Finance, the PLA's National Defense Science and Technology Commission (NDSTC), the State Council's National Defense Industry Office (NDIO), and the Central Military Commission's Science and Technology Equipment Commission (STECO).

In 1981 the 8th MBI (missiles) was merged into the 7th MBI (space), but this was largely a cosmetic move. The first major reorganization of China's defense-industrial complex occurred in 1982. At that time most of the numbered industries, which formerly belonged to the Ministry of Machine Building Industry (MBI), were reorganized into separate and distinct ministries. Five of them were defense related: the Ministry of Nuclear Energy, the Ministry of Aviation Industry, the Ministry of Electronics Industry, the Ministry of Ordnance Industry, and the Ministry of Space Industry. The 6th MBI was converted into a state-owned company, the China State Shipbuilding Corporation (CSSC). In addition, 1982 also saw the combination of NDSTC, NDIO, and STECO into a single entity called the Commission on Science, Technology, and Industry for National Defense (COSTIND).

In 1988 a reorganization that involved consolidating ministries took place: The Ministry of Nuclear Energy was combined with the Ministry of Coal Industry and Ministry of Electric Power to form the Ministry of Energy Resources; the Ministry of Aviation Industry and Ministry of Space Industry were combined to form the Ministry of Aerospace; and the Ministry of Electronics Industry and Ministry of Ordnance Industry were combined to form the Ministry of Machine Building and Electronics.

In 1993 another reorganization took place, this time consisting of redividing the ministries and converting some of the resultant entities into general companies *(zong gongsi)*: The nuclear energy portion of the Ministry of Energy Resources was converted into the China National Nuclear Corporation, and the Ministry of Coal Industry and Ministry of Electric Power Industry were reestablished; the Ministry of Aerospace was converted into two companies, Aviation Industries of China and China Aerospace Corporation; and the Ministry of Machine Building and Electronics Industry was broken down into the Ministry of Electronics Industry, Ministry of Machine Industry, and Northern Chinese Industries Corporation (NORINCO), the last of which comprised the enterprises formerly under the Ministry of Ordnance Industry. Thereafter, China's defense industrial sector officially consisted of five corporations: China National Nuclear Corporation, Aviation Industries of China, China Aerospace Corporation, NORINCO, and China State Shipbuilding Corporation.[26]

The goals of these reorganizations were to reduce enterprise reliance on government support, to spur economic dynamism, and to encourage innovation. These reorganizations (and the goals they embodied) were broadly consistent with, although weaker than, government policies toward all China's state-owned enterprises (SOEs), which aimed to make SOEs less dependent on state funds, more efficient, and eventually profitable and self-sustaining.[27]

As the frequency of the divisions and recombinations shown in Figure 5.1 suggests, in many cases these changes were largely cosmetic. Ministries that nominally were subdivided remained closely bound together; ministries that were officially combined generally remained organizationally distinct; and ministries that were converted into companies continued to be controlled by the government and behave like government ministries.[28] In the reform of civilian ministries in China, the depth of such reorganizations could typically be seen through their

---

[26] The best discussion of these organizational changes is in Frankenstein and Gill, 1996, pp. 398–400, but see also Gallagher, 1987; Frieman, 1989; Ostrov, 1991; Frankenstein, 1993; Frieman, 1993; Ding, 1996b; Gill and Henley, 1996.

[27] Naughton, 1996; Frankenstein and Gill, 1996; Brömmelhörster and Frankenstein, 1997.

[28] Kuan Cha-chia, "Jiang Zemin Sets Up General Equipment Department, Zhu Rongji Advances Military Reform," *Kuang Chiao Ching*, 16 April 1998, pp. 10–12, in FBIS as "Establishment of Military Department Noted," 6 May 1998; Yi Jan [sic], "Jiang-Zhu Relationship As Viewed From Army Structural Adjustment," *Ching Pao*, 1 March 1999, pp. 34–35, in FBIS as "Jiang-Zhu Relations in Army Reform Viewed," 9 March 1999; Li Xiuwei, "Applying Technology to National Defense," *China Space News*, 26 May 1999, p. 1, in FBIS as "Applying Technology to National Defense," 10 June 1999; Peng Kai-lei, "Five Major Military Industry Corporations Formally Reorganized," op. cit.; Xu Sailu, Gu Xianguang, Xu Xiangmin, *Zhongguo Junshi Kexue*, 20 June 2000, pp. 66–73, in FBIS as "Article on Effects of WTO Membership on PRC Military Economy," 27 July 2000.

Figure 5.1
Organizational Structure of Chinese Defense Industries

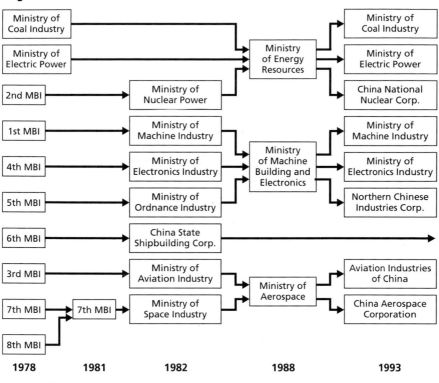

MBI = Machine Building Industry.
RAND *MG260-5.1*

physical effects on the headquarters of the entities in question. The headquarters of entities that were originally in separate locations often remained in their former locations even after being combined with another entity. When an entity was split, new headquarters for the two resulting enterprises were created by designating one half of the old headquarters compound as the headquarters of one of the new entities and the other half of the compound for the other. In some instances, the only apparent physical change was new nameplates on the compound gate. The story was no different for the ministries. When converted into corporations, ministers were redesignated general managers, but otherwise the personnel and organization were often unchanged.[29] The story in the defense sector was probably the same.

Explaining the Soviet Paradox[30]

A perennial question that arises when attempting to explain the weaknesses of China's defense industrial base is the "Soviet paradox": Why were the Soviets, despite all their economic failings, able to produce large quantities of relatively sophisticated weapons systems while the Chinese have not? This question is even more curious in that when China's defense industry was being rebuilt following the Civil War and the founding of the PRC, Chinese planners relied on assistance, organizational plans, and production processes from the Soviet Union, provided during their alliance in the 1950s. China's defense industry was built on the basis of the Soviet Union's industrial planning system, yet the quality and capability of the weapons produced by the two defense industries differed drastically.

Several factors explain the Soviet Union's relative success and China's failure with serial production of advanced military weapons. Historical considerations play a major role. The Soviet Union's industrial, scientific, and technical base was much broader and more developed than China's in 1949, when the PRC was established. The Soviets had made great strides in industrializing their largely agrarian economy since the 1920s. World War II resulted in a huge expansion of the Soviet defense industry and the bureaucratic structure to support it. Massive resources had been spent during the war to develop technologies and invest in personnel. China, by contrast, had little in the way of a defense industry (or industry of any kind) at the time of Japan's invasion in 1937. Japan's rapid occupation of China's most economically developed regions precluded significant development of the de-

---

[29] For a description of how this occurred in the petroleum sector, see Lieberthal and Oksenberg, 1988, pp. 123–127.

[30] This discussion is largely drawn from an unpublished manuscript by Peter Almquist entitled "Chinese and Russian Defense Industries: Problems and Prospects," written in 2003.

fense industry in China, as compared with the Soviet Union, which was able to keep much of its industry capacity out of Germany's hands during World War II. After Japan's defeat in 1945, the Nationalist government's military was largely dependent on U.S. military equipment. Thus, when the Chinese Communists came to power in 1949, China had little in the way of a defense industry.

In the Soviet Union, the defense industries were recognized for the key role they had played in winning World War II. China's experience following the end of World War II and its civil war could not have been more different. The success of the Chinese Communist Party had little to do with the capabilities of its weaponry: Mao's military strategies and overall strategic disposition, commonly referred to as People's War, stressed the primacy of man over machine. Consequently, Chinese leaders did not share the same political commitment to defense industrial production as the Soviet leadership did.

Another important historical difference between the Soviet Union and China is the timing of large-scale social and economic upheavals. The Soviet Union was in domestic turmoil between the start of World War I as a result of collectivization and the purges of the 1930s. However, by the end of the war, the Soviet defense industries had acquired some stability as the KGB gave some thought before arresting key engineers and designers in defense enterprises. In contrast, China went through the Great Leap Forward and Great Proletarian Cultural Revolution (GPCR) from 1958 to 1962 and from 1966 to 1969, respectively. This period of extreme societal and economic tumult devastated China's cadre of defense industry designers and technicians. The GPCR, in particular, destroyed the careers of many top scientists. While certain strategic weapons programs such as nuclear weapons and ballistic missiles were protected, the effect on the capabilities of China's other defense sectors and programs (such as the aviation industry) was significant. In contrast, after the death of Stalin, Soviet weapons designers and engineers no longer faced purges. By the mid-1960s, the Soviet Union had developed an impressive Chief Designer program and was churning out increasingly advanced weapons platforms.

During the GPCR, Mao initiated his policy of developing "third-line" defense production facilities. This called for the relocation and duplication of defense production facilities to China's interior provinces as a defense against a possible invasion by the United States or the Soviet Union. This decision greatly exacerbated redundancy and inefficiency in China's defense industrial establishment. Many of these interior locations were far from raw material and component suppliers; had poorly developed transportation links; and, because they were far from existing population centers, had little in the way of existing hospitals, schools, housing, and other facilities. These facilities had to be constructed from scratch. These problems were amplified by the practice of dispersing enterprises across multiple locations. Consequently, even communications and movement of materials within individual enterprises were hindered.

Differences in Soviet and Chinese threat perceptions and decisions about resource allocation during the Cold War further affected defense industrial development in the two countries. When the Cold War began, Soviet leaders perceived themselves as locked in a life-or-death competition with the United States. As a result, they continued to pour resources into the defense industry and to demand the utmost from that system. Chinese leaders did not feel the same urgency and, for several decades, were willing to accept secondary status in defense production to the Soviet Union. In the 1950s, Chinese leaders were initially concerned about consolidating the gains of the revolution, carrying out Mao's goal of the "communization" of agriculture, and transforming China's industrial sector through "the Great Leap Forward" campaign. During this period, China entered into many large defense trade agreements with the Soviets to build up its defense industries. In the 1960s, the Chinese leadership was preoccupied mainly with the Cultural Revolution, not defense. And even when reforms were initiated in the late 1970s, military modernization and defense industrial production were the last priority in Deng's four modernizations program.

The top leadership of the Soviet Union was heavily focused on the defense industries; many of the Soviet leaders had worked their way up the Party hierarchy in defense industrial enterprises and ministries. They were familiar with weapons production and had a strong proclivity to provide resources to the people and institutions that had backed their rise to power. Consequently, many of the decisions of the Party and State were geared toward promoting the defense industry. In China, none of the PRC's top leaders in the Politburo Standing Committee had strong personal or political ties to the defense industry. Although certain defense industry scientists, such as Nie Rongzhun, Zhu Guangya, and Qian Xuesen, were widely respected and influential in China, they never became senior leaders with the power to make resource allocation decisions directly.

Perhaps the most important reason the performance of the Chinese defense industry was so much poorer than that of its Soviet counterpart during the Cold War was China's lack of human and material resources. The Soviet Union had a far greater stock of scientists, engineers, and skilled workers than China did. It was also much richer. By the 1950s, the Soviet Union had a large and highly developed industry—civilian and defense—that was among the most advanced in the world in some areas of technology. The Soviet Union also had an extensive educational and research system and a literate, fairly well educated population. China, by contrast, had little in the way of industry and a minuscule cadre of scientists and engineers on which to build its defense industries. In 1952, less than one-half of one percent of the adult population of China had received a degree from an institution of higher education.[31] Thus, not only did China never devote as great a proportion of its human and material resources to its defense industry as the Soviet Union, it also had many fewer resources available during much of the Cold War. Only in the 1980s, when military spending was drastically reduced, did the size of China's industry and educated population begin to approach those of the Soviet Union.

---

[31] *China Statistical Yearbook 1996*, pp. 72, 632. Graduation rates for years prior to 1952 are not available but are assumed to be on average no higher than in 1952. The graduation rate for 1952 was estimated by dividing the number of graduates by the number of 65-year-olds (i.e., the population cohort that was born in 1930) in China in 1995 as provided in the above source. The resulting ratio (0.4 percent) is an upper bound for the graduation rate in 1952 because the size of this cohort was undoubtedly much larger in 1952 than in 1995.

New Progress in Defense Reform: The 1998 Defense Industry Restructuring and Beyond

Despite the heavy reliance by the Chinese leadership on organizational shuffling and reshuffling, reform measures prior to 1998 achieved little, if any, success in improving the performance of China's defense industry. Beijing had avoided implementing the major changes needed to create a dynamic defense sector. The government had done little to change the defense procurement system, which was plagued by inefficiency and corruption. Beginning in spring 1998, however, during the 9th Meeting of the National People's Congress, China's leadership initiated a new series of policies to reform the operation of the defense procurement system at the government level and to restructure the defense industries at the enterprise level. These policies have set the stage for institutional changes in the management of China's defense industry that outstrip past efforts in both scope and depth. These reforms indicate an acknowledgement of the depth of the problems of China's defense industrial system. They include policies intended to genuinely transform the structure and operation of that system by seeking to streamline and reduce corruption and inefficiency in the procurement process and forcing a degree of rationalization and accountability at the enterprise level. Assessing the current and future potential of China's defense industry requires a close examination of these reforms.

Beijing's New Strategy for Improving Defense Industrial Capabilities

Beijing's overall strategy for improving the technological capabilities of China's defense industries has three main elements. The first is selective modernization. China's leaders realize that given the size of China's economy and the overall technological level of the country, it would be too costly to attempt to acquire the capability to produce advanced weapon systems across the board. They observed how the Soviet Union's attempt to do so became a drag on the nation's economic development and are determined to avoid a similar fate. China's leaders are focusing on making breakthroughs in certain key areas.[32] These areas are

---

[32] Xiao Yusheng, Chen Yu, "Historic Leaps in China's Military Scientific Study," *Renmin Ribao*, 25 February 1999, p. 9, in FBIS as "Military Scientific Studies Take Leap," 2 March 1999;

rarely explicitly identified, but one article speaks of exploiting China's strength in aerospace, the manufacturing of missiles, and electronics technologies, while another advocates concentrating on "C4ISR, accurate strike weapons, and other crucial high-tech equipment."[33]

The second element of the strategy is civil-military integration. Despite the difficulties associated with defense conversion described above, China's leaders remain convinced that the integration of civilian and military production is the key to developing an advanced military. Although in the early 1980s the primary hope was that China's defense manufacturers would be able to use their technological capabilities to generate profits on civilian markets, today the principal hope seems to be that, through participation in commercial production, China's defense manufacturers will acquire dual-use technological capabilities that can be used in the production of weapon systems. China's leaders also continue to count on civilian production by China's defense manufacturers to maintain their financial solvency, reducing the amount of

---

An Weiping, "Thoughts on Developing Armaments by Leaps and Bounds," *Xinhua Domestic Service,* 1 July 1999, in FBIS as "Jiang Congratulates Defense Enterprise Restructuring," 2 July 1999; Gong Fangling, "There Should Be New Ideas in Defense Economic Building," *Jiefangjun Bao,* 14 September 1999, p. 6, in FBIS as "Article on 'Defense Economic Building,'" 23 September 1999; Li Xuanqing, Fan Juwei, Fu Mingyi, "All-Army Weaponry Work Conference Convened in Beijing," *Jiefangjun Bao,* 4 November 1999, p. 1, in FBIS as "Army Weaponry Work Conference Opens," 10 November 1999; Wang Congbiao, "Implement the Strategy of Strengthening the Military Through Science and Technology to Improve the Defensive Combat Capabilities of China's Military—Studying Jiang Zemin's 'On Science and Technology'," *Jiefangjun Bao* (internet version), 13 February 2001, p. 1, in FBIS as "Review of Jiang Zemin's Views on High-Tech Military," 13 February 2001.

[33] Maj. Gen Xun Zhenjiang, Captain Geng Haijun, "Exploring the Chinese Way to Develop Military Weaponry," *Zhongguo Junshi Kexue,* 20 June 2002, pp. 50–55, in FBIS as "Chinese General Recommends R&D Strategy for Weapons and Equipment," 20 June 2002; Zhang Zhaozhong, "Master New Development Trends of Military Equipment," *Jiefangjun Bao,* 14 April 1998, p. 6. The authors of the former article also express the hope that the ongoing "revolution in military affairs" will enable China to shortcut the process of developing an advanced military by achieving the "informationization" of its military at the same time as it accomplishes the still incomplete process of mechanizing the PLA. The author of the second article strongly opposes reverse engineering and copy production ("studied imitation") as a means for advancing China's military technology, as such an approach would leave China in a position of perpetually lagging behind the most advanced military powers.

funding the government needs to provide simply to keep these enterprises afloat.[34]

The third element of Beijing's strategy for improving the technological capabilities of China's defense industries is acquiring advanced foreign technology.[35] With the aim of not undercutting the long-term goal of self-reliance in defense production, China sees importing foreign technology as essential to enabling it to achieve independence in defense production. Given the backwardness of China's defense industries relative to the advanced nations of the world, Chinese leaders believe that the best way to rapidly achieve this goal is to import both the technology for producing state-of-the-art military equipment and the equipment itself until Chinese manufacturers can produce more technologically advanced products. As two Chinese military officers involved in defense production put it, China should "obtain jade from the rocks of other mountains" (他山之石可以攻玉), meaning that China should "learn or buy anything we can from foreigners" and "study and buy things by hook and by crook."[36]

---

[34] E.g., see Liu Jibin, "Implement the Guideline of Military-Civilian Integration, Rejuvenate the National Defense Science and Technology Industry," op. cit.; Liu Zhenying, Sun Jie, *Xinhua Domestic Service,* 1 July 1999, in FBIS as "More on Zhu at Defense Group Ceremony," 1 July 1999; *Central Television Program One Network,* 1 July 1999, in FBIS as "Zhu at Defense Ceremony," 1 July 1999; Peng Kai-lei, "Five Major Military Industry Corporations Formally Reorganized," *Xinhua,* 1 July 1999, in FBIS as "Zhu Rongji Urges Sci-Tech Work for National Defense," 1 July 1999; Ye Weiping, "Challenges and Opportunities for Ordnance Industry Following China's Entry to WTO (Part 2 of 2)," *Ta Kung Pao* (internet version), 26 April 2000, in FBIS as "Part 2: Ta Kung Pao on WTO Impact on Ordnance Industries," 3 May 2000; Wang Congbiao, "Implement the Strategy of Strengthening the Military Through Science and Technology to Improve the Defensive Combat Capabilities of China's Military—Studying Jiang Zemin's 'On Science and Technology,'" op. cit.

[35] An Weiping, "Thoughts on Developing Armaments by Leaps and Bounds," op. cit.; Tang Hua, "Science, Technology, and Industry for National Defense Increases Intensity of Innovation," op. cit.; Li Xuanqing, Fan Juwei, Fu Mingyi, "All-Army Weaponry Work Conference Convened in Beijing," op. cit.; Wang Congbiao, "Implement the Strategy of Strengthening the Military Through Science and Technology to Improve the Defensive Combat Capabilities of China's Military—Studying Jiang Zemin's 'On Science and Technology,'" op. cit.; Liu Jibin, "Implementing Thinking on 'Three Represents', Reinvigorate National Defense Science, Technology, and Industry," *Renmin Ribao* (internet version), 29 September 2001, p. 5, in FBIS as "Renmin Ribao on Implementing 'Three Represents' to Reinvigorate National Defense," 29 September 2001.

[36] Maj. Gen Xun Zhenjiang, Captain Geng Haijun, "Exploring the Chinese Way to Develop Military Weaponry," op. cit. See also General Li Xinliang, "Hi-Tech Local Wars' Basic Requirements for Army Building," *Zhongguo Junshi Kexue,* 20 November 1998, pp. 15–20, in FBIS as "Li Xinliang on High-Tech Local War," 17 May 1999.

This does indeed appear to be Beijing's strategy. Russian experts are currently providing workers at the Shenyang Aircraft Corporation with the know-how to assemble Su-27 fighter aircraft using imported materials and equipment. They are also training Chinese workers and engineers to manufacture many key materials domestically. China has also received weapons-making know-how from Israel Aircraft Industries in the form of assistance in designing and producing its J-10 fighter, and from large numbers of Russian scientists who are said to be employed by China's defense industries.[37] In addition to importing know-how, China has also been importing the machinery needed to manufacture sophisticated weapons systems, including illegal imports of substantial quantities of nominally civilian machinery and materials that can be used in the manufacture of weapons. China has imported from the United States machinery that can be used to produce materials that it cannot indigenously produce or legally import.[38] China has also been active in espionage activities to acquire knowledge needed to supplement indigenous R&D efforts.

Despite these comprehensive efforts to import foreign technologies, the determining factor in China's ability to produce advanced weapon systems will be the indigenous capabilities of its defense industries. Export controls and the efforts of foreign defense firms to maintain their competitive advantage will prevent China from being able to acquire the complete range of machinery, materials, and know-how needed to produce the advanced weapon systems the PLA desires. China will have to rely on the ability of its defense industries to develop many of the materials, machinery, and much of the know-how it needs and to integrate them with imported technology to produce complete, technologically capable weapon systems.

The Goals of the 1998–1999 Reforms

The reforms implemented in 1998–1999 were designed to achieve a number of broad goals related to further reforming the organization

---

[37] E.g., see Tung Yi, *Sing Tao Jih Pao*, 6 September 2000, p. A39, in FBIS as "Russian Experts Said Helping PRC Make High-Tech Weaponry," 6 September 2000.

[38] E.g., see U.S. General Accounting Office, 1996a, b.

and incentives of the defense industry establishment.[39] One goal was to further separate the state from enterprise operations. Although major defense firms had been converted from ministries into corporations in 1993 (1982 in the case of China State Shipbuilding Corporation), China's five defense companies had continued to behave very. much like the ministries from which they had been created. These companies were not only involved in production but also in regulatory and policy-making issues. The tensions and conflicts of interest resulting from this model created major impediments to making defense industry firms more efficient.[40]

A second goal was to introduce a "moderate" level of competition into the defense sector.[41] Because all the enterprises in each of the defense sectors were controlled by a single corporation, improvements in technology and management practices occurred more slowly than they might have in an environment in which multiple companies were striving to outdo each other.

---

[39] Premier Zhu Rongji articulated five goals following the initiation of the 1998 reforms: to separate state and enterprise functions; to establish a mechanism for "moderate" competition; to concentrate science and technology resources on weapons development and production; to promote better military-industry layout and restructuring; and to press enterprises to reduce their losses by helping to create a positive business environment to free enterprises from current difficulties. Tang Hua, "Science, Technology and Industry for National Defense Increases Intensity of Innovation," op. cit.; "Chinese Premier Underlines Science, Technology for National Defense," *Peoples Daily* (English edition), 2 July 1999; Peng Kai-le, "Five Major Military Industry Corporations Formally Reorganized," op. cit.

[40] Kuan Cha-chia, "Jiang Zemin Sets Up General Equipment Department, Zhu Rongji Advances Military Reform," op. cit.; Hsiao Cheng-chin, "Liu Jibin, Minister in Charge of the State Commission of Science, Technology, and Industry for National Defense and a Veteran Who Has Rejoined His Original Unit," op. cit.; Yi Jan, "Jiang-Zhu Relationship As Viewed From Army Structural Adjustment"; *Xinhua Domestic Service*, in FBIS as "Wu Bangguo Speaks at Defense Industry Conference," op. cit.; Li Xiuwei, "Applying Technology to National Defense," op. cit.; Peng Kai-lei, "Five Major Military Industry Corporations Formally Reorganized," op. cit.; *Xinhua*, in FBIS as "Zhu Rongji Urges Sci-Tech Work for National Defense," op. cit.; Tang Hua, "Science, Technology, and Industry for National Defense Increases Intensity of Innovation," op. cit.; Xu Sailu, Gu Xianguang, Xu Xiangmin, in FBIS as "Article on Effects of WTO Membership on PRC Military Economy," op. cit.; Xiao Yusheng, "Building a Strong People's Army," *Liaowang*, 29 July 2002, pp. 7–9, in FBIS as "PRC Article on PLA Military Buildup Over Last 10 Years, Preparations for Future," 8 August 2002.

[41] *Xinhua*, in FBIS as "Wu Bangguo Speaks at Defense Industry Conference," op. cit.; Liu Zhenying, Sun Jie, in FBIS as "More on Zhu at Defense Group Ceremony," op. cit.; *Central Television Program One Network*, in FBIS as "Zhu at Defense Ceremony," op. cit.; *Xinhua*, in FBIS as "Zhu Rongji Urges Sci-Tech Work for National Defense," op. cit.

Related to this, a third goal of the reforms was to provide more autonomy for individual enterprises within each defense corporation. The persistence of the ministerial system of organization meant that, while individual enterprises were nominally independent subsidiaries of the five defense companies, in practice the relationship between enterprises resembled that of organizations within a bureaucracy, with subordinate enterprises having little autonomy in their decisionmaking and internal management.[42] This stifled initiative and creativity.

At the same time as the reforms gave enterprises more autonomy, they also sought to make the various defense enterprises responsible for their bottom lines.[43] Like many of China's state-owned enterprises, defense enterprises that suffered excessive losses had not been penalized. Loss-making enterprises were not allowed to go bankrupt but were simply given subsidies or bank loans to make up the difference between revenues and expenditures. This lack of financial accountability not only discouraged defense enterprises from taking steps to cut losses but also provided no incentive to innovate, because the survival of an enterprise was unrelated to the quality of its products.

Paradoxically, while providing enterprises with more autonomy, the reforms also sought to increase the degree of horizontal and vertical integration within each of the defense sectors. Despite their hierarchical organization, the defense enterprises tended to operate in isolation, with little coordination or information sharing among them.[44] In par-

---

[42] Liu Zhenying, Sun Jie, in FBIS as "More on Zhu at Defense Group Ceremony," op. cit.; *Central Television Program One Network,* in FBIS as "Zhu at Defense Ceremony," op. cit.; Peng Kai-lei, "Five Major Military Industry Corporations Formally Reorganized," op. cit.; *Xinhua,* in FBIS as "Zhu Rongji Urges Sci-Tech Work for National Defense," op. cit.

[43] Liu Zhenying, Sun Jie, in FBIS as "More on Zhu at Defense Group Ceremony," op. cit.; *Central Television Program One Network,* in FBIS as "Zhu at Defense Ceremony," op. cit.; Peng Kai-lei, "Five Major Military Industry Corporations Formally Reorganized," op. cit.; *Xinhua,* in FBIS as "Zhu Rongji Urges Sci-Tech Work for National Defense," op. cit.

[44] Kuan Cha-chia, "Jiang Zemin Sets Up General Equipment Department, Zhu Rongji Advances Military Reform," op. cit.; Liu Jibin, "Implement the Guideline of Military-Civilian Integration, Rejuvenate the National Defense Science and Technology Industry," op. cit.; Si Yanwen, Chen Wanjun, *Xinhua Domestic Service,* 9 June 1999, in FBIS as "General Armaments Director on Developing Weapons," 9 June 1999; Xu Penghang, "Give Play to the Strength of Military Industries To Participate in Development of China's West," *Renmin Ribao Overseas*

ticular, the reforms sought to combine the functions of research and production, which traditionally had been carried out separately.[45]

Overall, the Chinese leadership's aim was to reshape the entire defense industry into three types of enterprises: "backbone enterprises" that would focus on military production, enterprises that would produce both military and civilian items, and enterprises that would focus on civilian production while using their technological capabilities to raise the overall level of China's science and technology base.[46]

Organizational Reforms

Beginning in 1998, Beijing adopted a series of policies to overhaul the organization and operation of China's defense industry. Reforms occurred both at the central government level and at the enterprise level. In general terms, the reforms aimed to centralize and standardize weapons procurement decisions at the central government level of operations while decentralizing defense enterprise operations in order to increase incentives for efficiency and innovation.

Central Government Reforms. The government adopted two major reforms that significantly changed the weapons procurement process. First, during the 9th National People's Congress meeting, the government abolished the military-controlled Commission on Science, Technology, and Industry for National Defense (COSTIND), which had been created in 1982, and replaced it with a strictly civilian agency of the same name but under the control of the State Council and then-premier Zhu Rongji. The new State COSTIND, which is run by civilian personnel, was formed by combining the defense offices of the Ministry of Finance, the State Planning Commission,

---

Edition (internet version), 24 March 2000, p. 2, in FBIS as "RMRB on Utilizing Military Industries To Develop China's West," 24 March 2000; Xu Sailu, Gu Xianguang, Xu Xiangmin, in FBIS as "Article on Effects of WTO Membership on PRC Military Economy," op. cit.

[45] *Xinhua Domestic Service,* in FBIS as "Wu Bangguo Speaks at Defense Industry Conference"; Tang Hua, "Science, Technology, and Industry for National Defense Increases Intensity of Innovation."

[46] Zhu, 1999, pp. 163–167.

and the administrative offices of the five major defense corporations.[47] Previously, COSTIND had reported to both the State Council and the CMC and was staffed by both civilian and military personnel. It had been responsible for overseeing all aspects of China's defense industries, including involvement in the daily management of China's large defense firms. As a quasi-military agency, the old COSTIND had also been very heavily involved in decisions on R&D and the purchase of military equipment. It had acted as a bridge between the PLA and defense enterprises. In that role, COSTIND exerted heavy, and in some cases preponderant, influence over defense procurement decisions. COSTIND's leading role contributed to the inefficiency of the procurement process: The PLA was unable to acquire the weapons it needed, weapons were frequently delivered late, and they were often defective or of very poor quality.[48]

In terms of defense procurement and defense production, the restructured COSTIND's responsibilities, resources, and authority have been substantially circumscribed. It is no longer heavily involved in decisions on the acquisition of new military equipment or the direct management of defense industry enterprises.[49] In stark contrast to its previous incarnation, the new COSTIND controls no procurement funds and thus possesses much less influence over procurement decisions.[50] The new COSTIND is tasked with coordinating procurement negotiations between the uniformed military and defense industrial enterprises. In that role, COSTIND is used by the military to coordinate multiple bids from defense enterprises and to ensure contract

---

[47] *Xinhua,* 10 March 1998, in FBIS as "NPC Adopts Institutional Restructuring Plan," 03/10/1998; Tseng Hai-tao, "Jiang Zemin Pushes Forward Restructuring of Military Industry—Developments of State Commission of Science, Technology, and Industry for National Defense and Five Major Ordnance Corporations," *Kuang Chiao Ching,* 16 July 1998, pp. 18–20, in FBIS as "Journal on PRC Military-Industrial Reform," 28 July 1998.

[48] Gallagher, "China's Military Industrial Complex." Gallagher was an assistant army attaché in Beijing, and his account is based on participation in many negotiations with the Chinese in the late 1980s.

[49] Tang Hua, "Science, Technology, and Industry for National Defense Increases Intensity of Innovation."

[50] Conversations with GAD officials, Beijing, China, October 2002.

compliance by the enterprises. PLA officials have indicated that, because COSTIND is a government agency, they can trust it more than the production enterprises to ensure that contractual obligations are met. COSTIND also controls some R&D funds for basic and applied research, although other institutions now appear to have control over most defense R&D funds. COSTIND's funds are not used to directly finance weapons production.[51]

COSTIND was also stripped of responsibility for direct management of defense enterprise affairs. The restructured COSTIND is meant to function as the administrative and regulatory agency for China's major defense enterprises. Its responsibilities include formulating laws and regulations to govern science and technological development for national defense, organizing international exchanges, defense cooperation with and arms sales to other countries, and export control administration related to military exports. Thus, COSTIND takes care of the government functions of China's defense enterprises while leaving the enterprises to manage themselves, a process in which COSTIND used to be heavily involved. This change is intended to allow the enterprises to focus on business decisions regarding production, cost control, and profitability.[52]

COSTIND's role, however, continues to evolve and its responsibilities may go beyond purely administrative and regulatory affairs. According to Liu Jibin, Minister of COSTIND from 1998 to 2003,

---

[51] Conversations with PLA officials, Beijing, China, May 2003.

[52] "Ten Military Industry Corporations are Founded," *Zhongguo Hangtian* (China Aerospace), August 1999 as translated in FBIS, 1 August 1999; "Speech of Liu Jibin at COSTIND Working Meeting," *Zhongguo Hangkong Bao* [China Aviation News], 30 April 1999 as translated in FBIS, 20 April 1999; "Interview by Central People's Radio Network reporter Zhao Lianju: Work Earnestly to Usher in the Spring of Science and Industry for National Defense—Interviewing Liu Jibin, Minister in Charge of the Commission of Science, Technology, and Industry for National Defense," 30 March 1998, in FBIS as "PRC Minister on Future Projects for Defense Commission," 30 March 1998; Gao Jiquan, "Shoulder Heavy Responsibilities, Accept New Challenges—Interviewing Liu Jibin, Newly Appointed State Commission of Science, Technology, and Industry for National Defense Minister," *Jiefangjun Bao*, 9 April 1998, p. 5, in FBIS as "New COSTIND Minister Interviewed," 9 April 1998; "National News Hookup," *China Central Television One*, 21 April 1998, in FBIS as "Interview with Minister of National Defense Science," 21 April 1998. Also see Tseng Hai-tao, "Jiang Zemin Pushes Forward Restructuring of Military Industry—Developments of State Commission of Science, Technology, and Industry for National Defense and Five Major Ordnance Corporations."

COSTIND's responsibilities and functions include "supervising the management of science and technology for national defense" and drawing up, organizing, and coordinating the implementation of development plans for the industry.[53] What precisely this supervision, organization, and coordination entail is unclear and suggests that governmental and corporate functions are still entangled. As Chinese analysts put it in a 2000 article on the military economy, "China is likely to maintain her centralized and unified state management by employing planning, administrative, or legal means. Moreover, China is also likely to tackle certain special issues, which might come up, by employing such means or methods as administrative intervention."[54]

The second major organizational reform, following the "civilianization" of COSTIND, was the creation in April 1998 of a new general department of the PLA known as the General Armaments Department (GAD—总装备部). GAD assumed the old COSTIND's responsibilities for military procurement combined with the roles and missions of the General Equipment Bureau under the General Staff Department, as well as divisions from the General Logistics Department related to other military equipment and procurement. It was largely formed by appropriating the personnel that handled these activities from these organizations. The responsibilities of GAD include the life cycle management of the PLA's weapons systems (from R&D to retirement) and running China's testing, evaluation, and training bases.[55] In addition, mainly through its Science and Technology Committee, GAD plays

---

[53] "Interview by Central People's Radio Network reporter Zhao Lianju: Work Earnestly to Usher in the Spring of Science and Industry for National Defense"; Gao Jiquan, "Shoulder Heavy Responsibilities, Accept New Challenges"; "National News Hookup," *China Central Television One*. See also "Zhuyao Zhize" 主要职责 ("Primary Responsibilities"), COSTIND website (http://www.costind.gov.cn/htm/jgjj/zyzz.asp), accessed 19 July 2004; *Xinhua,* 10 November 2000, in FBIS as "China Sets Targets For Aviation Industry," 10 November 2000.

[54] Xu Sailu, Gu Xianguang, Xu Xiangmin, in FBIS as "Article on Effects of WTO Membership on PRC Military Economy," op. cit.

[55] Harlan Jencks, "The General Armaments Department," in James C. Mulvenon and Andrew N.D. Yang, *The People's Liberation Army as Organization: Reference Volume v1.0,* Santa Monica, CA: RAND Corporation, 2002.

a role in broad policy debates about military modernization, defense procurement, and arms control and nonproliferation issues.[56]

The significance of the "civilianization" of COSTIND and the creation of GAD is twofold. First, these policy changes centralized China's military procurement system. Previously, responsibilities for PLA purchases had been divided among numerous civilian and military organizations, each with distinct and often conflicting interests. COSTIND's former predominant influence was a particular problem. The major responsibilities for identifying the PLA's needs and fulfilling them with appropriate equipment are now all located in GAD with input from the General Staff Department and the service branches. The formation of GAD improved linkages between the R&D and production stages in the procurement cycle. Previously these steps had been separate, resulting in inefficiencies and disconnects between the design and production of weapons systems. GAD is now in charge of setting goals and priorities and providing funding for the entire production process—from R&D to testing and evaluation, production, management, and eventual retirement and replacement.[57]

Second, the 1998 reforms separated the builders from the buyers. This organizational change further rationalized the procurement system and aimed to reduce conflicts of interest and corruption. GAD represents the PLA interests whereas COSTIND, as a civilian agency, now mainly handles industrial planning and the administrative affairs of defense firms. Because GAD controls procurement funds, PLA interests now play a decisive role in production decisions. When the former COSTIND had authority over production decisions, industry interests dominated—even though COSTIND included representatives of both the PLA and the defense industries. In many cases, PLA purchases were driven more by the production capabilities of certain defense firms (whose interests were promoted by COSTIND) than by

---

[56] Two of the committee's most senior members, Zhu Guangya and General Qian Shaojun, afford the S&T Committee a very influential role in Chinese civil and military nuclear affairs. Harlan Jencks, "The General Armaments Department."

[57] Yun Shan, "General Equipment Department—Fourth PLA General Department," *Liaowang,* 25 May 1998, p. 30, in FBIS as "China: New PLA General Equipment Department,"

the needs of the PLA. Unsurprisingly, the change from past practices has produced numerous conflicts and much bureaucratic competition between GAD and the new COSTIND. These have complicated full reform of the procurement system.[58]

In addition to these large organizational reforms, the government also adopted policies specifically related to changing the weapons procurement process.[59] Although most of these policies are quite new and their effect on actual production output is not yet clear, their adoption and implementation indicate the seriousness of the central government's efforts to reform the military procurement process and, ultimately, to improve production capabilities. According to Chinese officials, these policy reforms are driven by several broad goals: standardizing and centralizing the procurement system for military goods, establishing a legally based procurement system to protect both the military's and enterprises' contractual rights and responsibilities, adopting a system of market competition with open contract bidding and negotiation for defense purchases, and improving the quality and professionalism of the personnel involved in weapons procurement.[60]

These reforms have taken several forms. First and foremost, the military has sought to create a system that will standardize, unify, and legalize the weapons procurement process. Since the first procurement

---

12 June 1998; Pai Chuan, "Command system of the Chinese Army," *Ching Pao,* 1 December 1998, pp. 40–42, in FBIS as "Overview of PLA Structure," 12 December 1998; Xiao Yusheng, "Building a Strong People's Army."

[58] Conversations with GAD officials, Beijing, China, October 2002.

[59] In addition to focusing on reform of the procurement process, the PLA has devoted equal energy to the reform of the system of management of weapons systems. In the past three years, an entire cottage industry of books and studies on weapons equipment management (*wuqi zhuangbei guanli* 武器装备管理) has emerged in China. These polices are aimed at improving the system through which the military utilizes and maintain its weapons throughout their life span. Some examples of research work on equipment management include Ci Shihai, *Budui Zhuangbei Guanli Gailun* 部队装备管理概论 [Army Equipment Management Theory], (Beijing, China: Junshi Kexue Chubanshe, 2001).

[60] Xie Dajun, "The Procurement and Supervision of the Manufacture of Foreign Armaments," *Xiandai Junshi,* 15 August 1999, pp. 52–54 as translated in FBIS, 15 August 1999; Liu Cheng, "Creating a New Situation in Weapons and Equipment Modernization Effort," *Jiefangjun Bao,* 14 October 2002 as translated in FBIS, 14 October 2002.

regulations were adopted in 1990, numerous laws and regulations have been promulgated, leading to much confusion in the military purchasing system. In October 2002, Jiang Zemin signed an order promulgating and implementing a new set of regulations on military equipment procurement (*Zhonghua Renmin Jiefangjun Zhuangbei Caigou Tiaoli* 中华人民解放军装备采购条例).[61] These new regulations are meant to standardize several aspects of the procurement system including procurement planning, specification of procurement methods, equipment procurement procedures, procurement contract procedures, executing contracts, and purchasing foreign equipment.[62]

The new regulations are also meant to accelerate the establishment of a competitive bidding system for PLA contracts, which was discussed in 1998 when GAD was formed.[63] As of 2000, however, Chinese analysts wrote that "China has not yet built a selection or purchase system commensurate with the usual practice on the international market . . . nor has China introduced a sound demonstration or decision making system for weapons or equipment purchases."[64] Under the new system, defense enterprises are meant to bid for some military contracts at "market prices," an unclear concept given the limited competition in China's heavily subsidized defense sector. According to Chinese reports, the government has already established a special procurement agency that will eventually unify all the military's purchasing. A *Xinhua* report explained the rationale for this move. "The main task of the reform is to change the purchasing mode and centralize the purchasing [of]

---

[61] In 1990, the Central Military Commission issued "Work Regulations for the Management of Weapons and Equipment." Since then, additional regulations have proliferated. Chinese media announced the promulgation of the new rules but have not made them publicly available. "Central Military Commission Chairman Jiang Zemin Signs Order Promulgating and Implementing Chinese People's Liberation Army Equipment Procurement Regulations," *Xinhua*, 1 November 2002.

[62] For research on China's past procurement processes see Ravinder Pal Singh, (eds.), *Arms Procurement Decision Making: China, India, Israel, Japan, South Korea and Thailand*, Stockholm International Peace Research Institute, (Oxford, United Kingdom: Oxford University Press, 1998.)

[63] "Government Procurement Again Recommended at NPC," *Xinhua*, 8 March 1999 as translated in FBIS, 8 March 1999.

[64] Xu Sailu, Gu Xianguang, Xu Xiangmin, in FBIS as "Article on Effects of WTO Membership on PRC Military Economy," op. cit.

items [which is] scattered in various departments to a degree as high as possible in the hands of an institution specializing in doing purchases . . . [t]hus gradually setting up a model based mainly on centralized purchasing."[65] This agency will be used for all the PLA's purchases except for highly confidential items, materials and equipment costing RMB 500,000 ($60,300) or more, and engineering projects costing RMB 2 million ($241,200) or more. In addition, projects covering a floor space of 2,000 square meters have to be carried out through public tender.[66] The use of this new procurement system is being piloted in specific PLA and PAP units until 2005, after which it may become universally adopted throughout the PLA.[67] The limited scope of the new agency and the new regulations (e.g., nothing over RMB 2 million) suggest that it is initially being used to centralize procurement of quartermaster-type goods such as uniforms but that it does not apply to major weapons systems—an area that needs to be revamped.

Enterprise-Level Reforms. Beginning in 1998, Beijing adopted far-reaching policies to alter the relationship between the government and defense enterprises. The central government's main goals were to separate the government administrative units from enterprise operations, to make the enterprises more market-oriented by exposing them to competitive pressures, to provide harder budget constraints, to make the enterprises less reliant on state subsidies, and to lessen the classic social burdens associated with the *danwei* (work unit) system. Chinese

---

[65] "PRC Plans Reform of Army Purchasing System," *Xinhua,* 9 January 2002; "PRC Armed Forces Adopt Government Procurement System to Meet Demands of Economic Reforms," *Xinhua,* 9 January 2002; "Central Military Commission Chairman Jiang Zemin Signs Order Promulgating and Implementing Chinese People's Liberation Army Equipment and Regulations," *Xinhua,* 1 November 2002.

[66] "Standardizing Our Military Armament Procurement Work According to Law," *Jiefangjun Bao,* 2 November 2002 as translated in FBIS, 2 November 2002; "PRC Plans Reform of Army Purchasing System," *Xinhua,* 9 January 2002 as translated in FBIS, 9 January 2002; "PRC Armed Forces Adopt Government Procurement System to Meet Demands of Economic Reforms," *Xinhua,* 9 January 2002 as translated in FBIS, 9 January 2002; "Central Military Commission Chairman Jiang Zemin Signs Order Promulgating and Implementing Chinese People's Liberation Army Equipment and Regulations," *Xinhua,* 1 November 2002 as noted in FBIS, 1 November 2002.

[67] "PRC Plans Reform of Army Purchasing System," *Xinhua,* op. cit.; "PRC Armed Forces Adopt Government Procurement System to Meet Demands of Economic Reforms," *Xinhua,* op. cit.

policymakers initiated several different types of reforms to change defense enterprise operations.

The major enterprise reform, which occurred in July 1999, involved the bifurcation of each of China's five core defense companies into two defense industrial enterprise groups (军工集团公司), or sometimes referred to as group corporations. An eleventh enterprise group, for defense electronics, was established in late 2002. The Chinese government pursued two goals in undertaking this reorganization. The obvious goal was to inject competition into the defense industrial sector. China's leaders hope that competition will cause defense enterprises to become more efficient, less of a financial burden on central and local governments, and more capable of independent innovation and absorbing and assimilating new technologies. The new head of COSTIND, Liu Jibin, explained in July 1999 the core rational for this organizational change. He noted that each of the general companies is divided into two roughly equivalent groups in terms of capability. It is expected that through "proper competition each of the new companies' efficiency can be improved and management mechanisms can be transformed."[68]

The extent to which real competition has emerged between the enterprise groups in each sector is not clear. As addressed in the case studies, in some sectors limited competition over defense systems, key subsystems, and parts has emerged. In others, the products of the two group companies are different enough that there is little to no direct competition between the two companies. Some Chinese writings indicate that the real goal of the recent reforms was not to promote competition in terms of products but rather in terms of "systems of organization" and "operational mechanisms."[69] That is, splitting each defense sector into two enterprise groups is apparently supposed to allow them to separately explore different approaches to organization and management so that they can learn from each other's successes and failures. This is an approach that might be adopted for increasing the efficiency of government organizations, not an effort to allow the defense enterprise groups to behave as true competitors operating in a free market. It again suggests

---

[68] "Ten Military Industry Corporations Are Founded," *Zhongguo Hangtian*, op. cit.

[69] Li Xiuwei, "Applying Technology to National Defense," op. cit.

that China's defense industries are still to some degree treated like government agencies, not as truly independent economic entities.[70]

The other goal of the 1999 bifurcation policy was the formation of "group corporations" (集团公司). This new category of company was an element of the government's broader effort to reform ownership structures in SOEs, including defense enterprises.[71] These group corporations were part of an effort to establish shareholder relationships within a company to further remove the government from firm operations, to distribute risk, and to increase accountability for profits and losses.[72] Many of China's major firms under the large group corporations have a long list of shareholders that own noncontrolling stakes in some operations. This effort at ownership reform is one of the newest policies aimed at orienting enterprise operations toward markets.

Beyond these basic structural reforms, Chinese policymakers have also implemented a variety of smaller initiatives to improve defense enterprise operations. First, enterprises are beginning to use nongovernment funds for military-related production projects. Enterprises manufacturing military equipment may use capital from other firms, internal monies, or investment from foreign countries to fund projects that produce weapons that are then marketed to the PLA or abroad. For example, the Chengdu Aircraft Corporation (AVIC I) is using its own and Pakistani funds to subsidize the development of the FC-1/Super-7 light fighter program. This platform is in the final stages of development, and the plane's first test flight recently took place. It is not clear whether the PLAAF is going to buy the plane, however. The FC-1 is primarily intended for the Pakistani and other third-world militaries, but it is possible that the PLAAF will acquire some. While the FC-1 does not represent a major technological leap for the Chinese aviation industry, its financing mechanism is innovative. Other defense firms,

---

[70] The fact that the headquarters of the new enterprise groups were apparently formed simply by subdividing the headquarters compounds of the old defense corporations reinforces this impression. See Peng Kai-lei, "Five Major Military Industry Corporations to Be Reorganized," op. cit.

[71] Lin and Zhi, 2001, pp. 305–341.

[72] Peng Kai-lei, "Five Major Military Industry Corporations Formally Reorganized," op. cit.

such as the Guizhou Aviation Corporation, which is developing an advanced jet trainer, have begun marketing designs of future systems in an effort to find investors to support the development of new weapons systems now that PLA procurement is no longer guaranteed.[73]

Second, some defense firms have created subsidiary organizations that have been listed on domestic Chinese stock exchanges in Shanghai or Shenzhen. These joint stock companies provide the controlling enterprise with an additional source of capital for reinvestment into enterprise operations. Listing on Chinese capital markets has the additional advantage of carrying with it minimal transparency requirements, an issue of concern to defense companies. To date, over 40 defense enterprises have been listed on the Shanghai and Shenzhen exchanges. Although some of these companies have been fully converted to nonmilitary production (but are still considered military companies by dint of their origins), some of the military companies listed continue to be involved in military projects as well as production of civilian goods.[74]

A third emerging change in the defense industry is the expansion of partnerships with civilian universities and research institutes to improve educational training relevant to the development of new military technologies. The growing number of partnerships between the defense industry and national educational institutions is notable. GAD and COSTIND have begun to partner with universities in various parts of the country to improve the PLA's access to individuals with technical training. In 2002, COSTIND gave several million renminbi to at least two aerospace and shipbuilding academies in Jiangsu Province that were not part of its past educational network, to develop their defense-related course offerings, to recruit students interested in defense research, and to give the staff of these academies additional training on defense technology issues.[75] These partnerships are with institutions other than the numerous universities that have traditionally been linked to China's defense industry establishment.[76] This gov-

---

[73] Data from Guizhou Aviation Corporation brochure, 2002.

[74] "Jungong Shanshi Qiye 2001 Nian Pandian," *Zhongguo Junzhuanmin* [China Defense Conversion], July 2002, p. 4.

[75] See *Guangming Ribao*, 27 August 2002; *Guangming Ribao*, 6 June 2002.

[76] COSTIND directly administers seven institutions of higher education: Harbin Institute of Technology; Beijing University of Aeronautics and Astronautics; Beijing Institute of Technol-

ernment-industry-university cooperation is most notable in the information technology (IT) field but is rapidly growing in other sectors involved in defense R&D and production.

Fourth, in recent years a limited amount of rationalization has occurred in the defense industry, although much more is needed in light of the continuing inefficiencies and redundancy still prevalent in the sector. Some shipbuilding and "ordnance" (ground systems) industrial groups, for example, have transferred ownership of large enterprises to provincial authorities in recent years. This has been a major trend in the shipbuilding industry where only 60 of China's shipyards are now controlled by the CSSC and the China Shipbuilding Industry Corporation (CSIC); the remainder are run by provincial and local authorities. Another aspect of rationalization has been layoffs and downsizing. This has been occurring in fits and starts. According to one report, 61,000 ordnance industry workers were laid off in 2001, and 100 other enterprises were earmarked for bankruptcy or takeover the following year.[77]

A fifth reform initiated by COSTIND and GAD is the promotion of R&D and production cooperation among defense enterprises located in various provinces. In the past, one of the organizational deficiencies of China's defense industries was the extensive reliance on single source suppliers in production of defense platforms. This problem has been particularly acute in China's aviation industries. Such practices have contributed to the inefficiency, redundancy, and high degree of insularity in the defense industry. GAD and COSTIND are trying to address this problem by promoting greater integration and information sharing among defense enterprises and R&D institutes in various provinces. During 2000, for example, the Beijing Military Representative Bureau reportedly began to cooperate with its counterparts in the national defense departments of universities, colleges, and scientific research institutes of five cities in northern China. The aim of this initiative was to facilitate better information sharing across prov-

---

ogy; Nanjing University of Aeronautics and Astronautics; Northwestern Polytechnical University; Nanjing University of Science & Technology; and Harbin Engineering University. In addition, the provincial branches of COSTIND, together with provincial governments, jointly administer a number of institutes and schools. See, for instance, http://www.jxgfgb.gov.cn/.

[77] "Defense Commission Minister Sets Targets for 2002," *Zhongguo Xinwen She,* 7 January 2002, in FBIS, 7 January 2002.

inces about technical innovations and potential markets for new products—both military and civilian.[78] While it is far from clear how successful this effort has been in overcoming deeply ingrained localism in defense production, the adoption of this plan to boost cross-province defense industry cooperation indicates that the government recognizes these problems and is making efforts to overcome them.

Last, the reform of the military representative office (MRO) system has become a recent priority for senior leaders in the GAD system. For years, the PLA used a system of military representative offices at the city, enterprise and factory level to ensure quality control and contract compliance of defense projects at factories and research institutes.[79] In light of the high degree of diversification of defense factories into civilian production since the 1980s, most MROs are based in civilian factories and institutes involved in defense work. The MRO system has been troubled and ineffective for many years. MROs are understaffed and military personnel are reportedly overworked. Many MRO personnel lack the technical expertise to effectively oversee contract compliance and ensure quality, a problem that in some cases is exacerbated by high staff turnover. Overall, staffers have done a poor job of monitoring and evaluating ongoing equipment production. The objectivity and reliability of many MROs are problematic because representatives who reside at factories for a long time tend to shift their loyalties from the military to protecting the interests of local factories and townships.[80]

These weaknesses in the MRO system are significant because many manufacturers show a disregard for ensuring the quality and reliability of their products and often miss production deadlines. Facto-

---

[78] Jiang Huai and Fu Cheng, "Beijing Military Representatives Bureau Cooperates with Five Provinces and Cities in North China in Building Regional Cooperation with Various Layers and Professions," *Jiefangjun Bao*, 18 September 2000, p. 8., as translated in FBIS, 18 September 2000.

[79] Chinese media reports have identified Military Representative offices in Beijing, Wuhan, Shenyang, Changsha, Shanghai, and Wuhan. According to one report there were seven military representatives in a factory. See *Jiefangjun Bao*, 14 November 2001, as translated in FBIS as "PRC: Article on PLA Plant Manufacturing Special Military Vehicles," 14 November 2001.

[80] The authors are grateful to Tai Ming Cheung for his insights on the MRO system; also see Wu Ruihu, "Navy Military Representative Hard at Work in Supervising Armament Development," *Jiefangjun Bao*, 10 April 2002 as translated in FBIS, 10 April 2004; "Military Representatives of Engineering Corps Work Hard to Ensure Assault Boats Quality," *Jiefangjun Bao*, 17 July 2002, p. 10 as translated in FBIS, 2004.

ries often give a higher priority to the production of civilian products than military ones due to their higher profit margins. Consequently, in recent years the government has initiated a major effort to overhaul the MRO system to improve contract compliance and quality control monitoring. Both the State Council and COSTIND issued a series of new policies on improving the quality of military production. These new regulations include an entire system for improving monitoring and boosting education of military representatives. The effectiveness of these measures, however, is far from clear.[81]

These efforts to improve the MRO system and to boost quality control are being adopted in parallel with new standards within GAD for training and utilizing staff with technical skills. GAD has adopted new measures to recruit, train, and retain personnel with science and technology training.[82] A report from China's *Science and Technology Daily* indicates that the specific GAD policies include "to actively recruit and replenish high quality talent, provide the talent with positions compatible with their skills, establish special positions in high priority disciplines, gather outstanding experts to serve armament development and research, and build up post-doctoral mobile stations into the frontline for recruiting high level staff. . . . "[83]

Constraints on China's Defense Industry Reform

The government's ability to reform the management of defense procurement and to change the incentive structures in and among defense enterprises will have a direct effect on the future production capabilities of China's defense industrial complex. Despite the many reforms

---

[81] Zhang Yi, "The State Adopts Effective Measures for Improving Quality of National Defense-related Products," *Xinhua,* 23 March 2000 as translated in FBIS, 23 March 2000; Fan Juwei, "Quality of Our Large-sized Complicated Armaments is Steadily Improving," *Jiefangjun Bao,* 19 July 2001, p. 1, as translated in FBIS, 19 July 2001.

[82] *Xinhua,* 11 April 2000, in FBIS as "PRC's PLA 'Speeds Up' Training for Armament Officers," 11 April 2000; Zou Fanggen, Fan Juwei, "Chinese Army's Armament, Scientific Research, and Procurement System Insists on Simultaneously Promoting Development of Scientific Research and Cultivation of Skilled Personnel," *Jiefangjun Bao* (internet version), 19 February 2002, p. 1, in FBIS as "PRC: PLA Implements Measures To Simultaneously Train Personnel, Develop New Weaponry," 19 February 2002.

[83] Liu Cheng, Jiang Hongyan, and Liu Xiaojun, "Talented Personnel to Support Leapfrog Developments of Weaponry," *Keji Ribao,* 31 October 2001, internet version, www.stdaily.com as translated in FBIS, 31 October 2001.

in institutions and incentives in the defense industries since 1998, it is not yet clear how efficacious these reforms have been. Beijing has a long and highly blemished history of adopting weak reforms and failing to implement more radical policy changes. It is not clear how quickly and effectively the post-1998 measures can overcome the inertia and extensive problems that have plagued China's defense industrial establishment for the past several decades. Many of these problems are deeply entrenched in the central government bureaucracy, provincial-level agencies, and enterprise-level business operations.

The government's success at fully implementing defense industrial reforms will be broadly influenced by several tensions or "contradictions" that persist at both the central government level and the enterprise level. These tensions constitute the broad, systemic constraints on China's newest, post-1998 effort to reform the defense industry system. The major tensions include the following:

Reform Imperatives Versus Social Stability. Efforts to rationalize and downsize China's large, bloated, and inefficient defense enterprises raise concerns about social stability, especially increasing unemployment, inability to fulfill pension commitments, and cutting off funding for enterprise-run social welfare programs. The 1998 reforms eliminated a number of the social welfare obligations of many state-owned defense enterprises, such as housing and health care. However, the managers of some defense enterprises have been reluctant to implement these reforms for fear of the impact on social stability. Riots and social unrest related to rationalization at defense factories have already occurred in China. Such concerns will likely limit the pace and scale of defense enterprise reform.

GAD Versus State COSTIND. The civilianization of COSTIND and the creation of GAD injected new tensions into the 1998 round of defense industry reforms. GAD gained influence over central government procurement-related decisions at the expense of COSTIND. These new agencies often compete for influence in the defense procurement process. This tension contributes to delays and inefficient decisionmaking on specific military projects. The competition between these agencies will continue to complicate the government's ability to streamline the procurement process and to reform defense enterprise operations.

Localism Versus "Free Market" Practices. Historically China's defense industrial enterprises were highly vertically integrated and relied on single-source suppliers. These patterns have been exacerbated by long-standing political ties within regions and provinces that influence business relations among firms in the same localities. As a consequence, many defense enterprises are reluctant to seek cooperation with firms in other regions, even though they may offer higher-quality and lower-cost products. This constitutes a significant barrier to improving the quality of weapon systems and reducing the costs of defense production.

Evaluation of the effect of the post-1998 series of reforms is constrained by the limited amount of data available on the actual operations of China's defense enterprises. Although the production capabilities of China's defense industry are most often assessed by examining the weapons and products produced by these plants, data and analysis of production processes have been limited. The following section offers a brief analysis of the aviation, missile, shipbuilding, and information technology and defense electronics sectors of China's defense industry. The case studies focus on the organization and operation of enterprises within these sectors. They are designed to illuminate how specific reforms have been implemented and to show their impact on the operations and output of China's defense sector.

Assessing Defense Industry Capabilities

The changes in the structure, operation, and production capabilities of China's defense industrial complex are most evident at the level of the individual industries. Below, we present the results of four case studies: aviation, shipbuilding, information technology and defense electronics, and missiles.[84] As with most aspects of modernization in China, reform of the defense industry has been uneven. We find that in each of these sectors the capabilities of manufacturers to design and produce key systems are improving, but weaknesses and limitations persist. Some sectors have been more successful than others: Improvements in

---

[84] More detailed analyses of these cases will be provided in a separate RAND study now in preparation.

information technology and shipbuilding have been very impressive whereas aviation has lagged.

Aviation Industry

China's military aviation industry is in the midst of a gradual transformation. It has made more progress in improving quality and the technological sophistication of aircraft in recent years than in previous decades. Before the late 1990s, all the fixed wing military aircraft produced in China, and their major subsystems, were essentially improved versions of 1950s Soviet technology. Since the late 1990s, China has begun producing progressively more advanced aircraft and subsystems. For example, except for the engines, which are imported, the JH-7 fighter-bomber is China's first completely indigenous combat aircraft design and the first third-generation fighter aircraft produced in China. China's aviation industry is now acquiring the capability to manufacture even more advanced fighter aircraft. The Russian-designed Su-27s being assembled in Shenyang, Liaoning Province are fourth-generation fighters; the Chengdu Aviation Corporation's J-10, which may now be entering production, is expected to be as well. The J-10 is primarily a domestic development (albeit designed with Israeli assistance). A similar story holds for major subsystems, such as engines. China has gone from producing improved versions of Soviet turbojets to the indigenously designed Kunlun turbojet engine, and it will soon be producing an indigenously developed high performance turbofan engine, the WS-10.[85]

These recent advances represent a noteworthy rate of improvement. In the space of a little over a decade, China's aviation industry has gone from producing second-generation aircraft to third-generation aircraft and is on the verge of producing domestically developed fourth-generation aircraft. This progress still leaves China a generation behind the United States, however.

In the area of rotary wing aircraft, China has shown less dramatic but nonetheless steady improvement in production capabilities. Since beginning license-production of the Eurocopter Dauphin 2 in the early

---

[85] Kenneth Munson, "CAC J-10," *Jane's All the World's Aircraft*, 22 April 2004, online at http://online.janes.com (accessed 18 May 2004); "Chinese Puzzle," *Jane's Defence Weekly*, 21 January 2004; Yihong Chang, "China Launches New Stealth Fighter Project," *Jane's Defence Weekly*, 11 December 2002; U.S. Department of Defense, 2004.

1980s, China's aviation industry has progressed to the point where it is now capable of domestically producing most or all the components for the Dauphin 2, two other Eurocopter designs, and indigenously developed variants of the Dauphin. China's aviation industry also produces the Arriel turboshaft engine under license. It now has the capability to produce relatively modern multirole helicopters and their engines. Although it took 20 years to get to this point and China has yet to produce an indigenously designed helicopter, a Chinese-designed multirole helicopter, the Z-10, is under development and could enter production sometime in the next few years.[86]

There remain important gaps in China's military aviation production capabilities that will not soon be filled. Until the development of the WS-10, China had been unable to successfully manufacture a turbofan engine. Because of this and other technological weaknesses, China produces no long-range heavy bombers or jet transports. Similarly, the immaturity of China's helicopter design and production capabilities has meant that China does not produce true attack helicopters (although the Eurocopter design has been modified to produce an attack variant). Programs to fill these gaps are still in their formative stages.

Much of China's recent progress in aviation production can be attributed to "latecomer's advantage." China draws heavily on advanced aviation technologies that have already been developed abroad. In recent years, it has been able to acquire expertise more rapidly and less expensively than if it had attempted to develop these technologies on its own. This approach is not without its drawbacks. Although Boeing, Airbus, General Electric, United Technologies' Pratt & Whitney unit,

---

[86] Kenneth Munson, "HAI (Eurocopter) Z-9 Haitun," *Jane's All the World's Aircraft*, 24 November 2003, online at http://online.janes.com, accessed 8 March 2004; Wang, "New Military Aircraft Displayed at the National-Day Grand Military Parade," *Liaowang*, 8 November 1999, pp. 30–31; Xu Dashan, *China Daily* (internet version), 13 September 2001, in FBIS as "China Daily: Helicopter Sector To Be Promoted," 13 September 2001; Xu Dashan, *China Daily* (internet version) 11 July 2002, in FBIS as "PRC's New H410A Helicopter Model Receives Certification; CAAC Says 'Huge Achievement,'" 11 July 2002; Robert Sae-Liu, "China Advances Helicopter Projects," *Jane's Defence Weekly*, 3 May 2002; Kenneth Munson, "CHAIG Z-11," *Jane's All the World's Aircraft*, 17 June 2003 (online at http://online.janes.com, accessed 8 March 2004); Kenneth Munson, "CHAIG Z-8," *Jane's All the World's Aircraft*, 17 June 2003 (online at http://online.janes.com—accessed 8 March 2004); Kenneth Munson, "CHRDI Z-10," *Jane's All the World's Aircraft*, 22 April 2004, online at http://online.janes.com, accessed 19 May 2004.

Rolls-Royce, and other Western manufacturers in the aviation industry have joint venture operations and subcontracting arrangements in China, thereby facilitating the acquisition of a range of production technologies by their Chinese partners, these partners must still master and assimilate those technologies. In addition, Western countries, with some exceptions, are not willing to transfer technologies to China that have direct military applications. Few, if any, foreign companies are willing to provide China with their most advanced "core" technologies, although Russian and Israeli companies appear to be willing to provide China with some advanced military technologies that U.S. or French companies would not.

As the capabilities of China's aviation industry begin to approach those of the rest of the world, the latecomer's advantage will no longer obtain. Other countries will be increasingly unwilling to provide China with more advanced aviation technologies. Moreover, it is unclear whether the Russian aviation industry, one of China's key suppliers, will have the technological capability and resources to create and manufacture significantly more sophisticated designs in the future. In the absence of an indigenous combat aircraft program, Israel is unlikely to be able to provide China with newer aviation technologies in areas other than subsystems such as avionics. Further improvements in the capabilities of Chinese aircraft will increasingly depend on domestic R&D and improvements in domestic production capabilities, i.e., China's indigenous capacity for technological progress.

Aside from limitations on its access to foreign technologies, China's aviation sector continues to face a number of other challenges. Some of the most talented individuals in the industry have left the state-owned aviation companies for joint ventures or private companies in China's coastal regions. The rapid development of the coastal areas has made settling in interior regions even less attractive for educated Chinese; recent college graduates are very reluctant to accept low-paying jobs in China's interior, where the aviation companies are located.

The impetus for technological progress provided by competition is still not much of a factor in China's aviation sector. Competition between the two large state-owned group companies or among their subordinate enterprises still appears to be minimal. This is true for component suppliers as well as the major airframe and engine manufacturers.

Lack of competition reduces the incentives for increased efficiency and innovation.

Organizational problems also persist. The aviation industry's research and design institutes remain organizationally and fiscally separate from the production enterprises. This separation limits flows of knowledge between the two sets of organizations and the incentives the institutes face to develop practical designs that the companies can manufacture. It is also unclear whether China has developed a network of training and technology consulting and service companies that would facilitate flows of knowledge through the industry.

Finally, China's capital markets still provide little pressure to improve efficiency. Although some aviation companies are now listed on Chinese stock markets, their core operations remain state-owned. State ownership provides little incentive for China's major military airframe and engine producers to pursue profits, which would stimulate them to become more efficient and innovative.

None of these shortcomings is unique to China. Even in the United States competition in the military aviation sector is constrained by the capital-intensive nature of the industry: Only two military airframe manufacturers remain—Boeing and Lockheed-Martin. Moreover, highly sophisticated aviation companies are at least partially state-owned in many countries. The Soviet Union built a formidable military aircraft industry even though its design institutes and production enterprises were also organizationally separate from each other and the Soviet aviation industry lacked a network of training and technology consulting and service companies. However, despite these similarities between China's aviation industry and those of other countries, China faces disadvantages that these countries do not. China lacks the highly developed economy and relatively unrestricted access to shared military technology that NATO members like France or Britain have. Similarly, while resources for military aviation have increased in recent years, in no way do they approach the level that the Soviet Union devoted to the aviation industry during the Cold War. Finally, more so than in any of these countries, China's aviation sector must compete for talented workers and managers with other industries that can offer positions that pay far more, are more prestigious, and promise a higher quality of life.

Thus, although the technological gap between China's military aviation industry and that of the United States and other major aviation producers will likely narrow in coming years, the gap will probably remain significant unless China makes fundamental changes in contracting and enterprise management. One such change would be the introduction of true competition in the form of open bidding for R&D and production contracts. Another would be the integration of design institutes with production enterprises.[87] A third would be the privatization of China's major military airframe and component manufacturers or, short of that, their listing on stock markets. Even if the government retained a controlling interest, partial privatization would likely have a positive impact because the pressure to achieve profits would drive improvements in efficiency and accelerate technological progress.

Shipbuilding Industry

China's shipbuilding industry (SBI) has made impressive strides since Beijing adopted its economic reform program in 1978. Over the last twenty years, China's shipbuilding industry has become the third-largest builder of commercial ships in the world, a major accomplishment given the minuscule level of commercial ship construction in the late 1970s. Prior to the reform period, most of China's shipbuilding was for military projects. Like many other heavy industries in China, at the beginning of the reform period shipyards leveraged their cheap costs of labor and materials to enter the low end of the global shipbuilding market. Chinese yards became known for producing inexpensive simple ships, although quality assurance and delays in delivery have been persistent problems. Since the 1980s, China has consistently expanded its share of the global shipbuilding market. As measured by dead weight tons (DWT), China now controls about 10 percent of the market.

---

[87] This process is already occurring to a certain extent in China. Some design institutes have been absorbed by manufacturing enterprises, although many have resisted integration; many manufacturing enterprises are establishing their own internal design institutes and test facilities; and some research institutes are expanding into production. Li Jiamo, The Design of Primary Aircraft Components Should Be Closely Integrated with Manufacturing Enterprises, *Guofang Keiji Gongye* [*National Defense Science and Technology Industry*], 2002, No. 5.

Over the past decade, major Chinese yards have upgraded their production capabilities and expanded their capacity. Technology-sharing agreements with foreign shipbuilders and adoption of foreign project management techniques have facilitated these processes. Chinese yards have adopted many modern production techniques, such as modular construction (as opposed to keel-up), which have made possible the serial production of new designs, especially naval platforms. The time needed to produce new major platforms has diminished substantially in recent years. SBI research and design institutes have also made great efforts to adopt and assimilate modern design techniques. Most of these design bureaus are linked to the actual shipyards, which has improved the overall process of designing and constructing ships. In other parts of China's defense industry, the links between R&D and production have not been as strong.

The shipbuilding industry was recently reorganized to stimulate competition among yards and encourage the larger shipyards to specialize. China's major shipyards regularly compete with each other for orders from foreign ship buyers; the government plays a limited role in this process. A bevy of large and modern shipyards are now coming on line in China. Some of them have begun to focus on building specific ship types, allowing a degree of specialization to emerge in the industry. Over time these changes should contribute to additional improvements in efficiency and profitability. The new and expanding interactions between R&D institutes and academic organizations with the industry should lead to further improvements in the SBI's research, development, and design capabilities.

These advances in design and construction capabilities have benefited Chinese naval projects—in great part as a result of the high degree of colocation of merchant and naval shipbuilding. Improvements in civilian shipbuilding capabilities have helped remedy many of the basic problems exhibited by China's first- and second-generation naval combatants, such as poorly designed and fabricated ships. Chinese destroyers and submarines built in the late 1990s and early 2000s exhibit significant improvements in design and construction compared with the previous generation. The newest vessels are more seaworthy and battle-ready; the newest designs permit the inclusion of more modern weapons and sensor suites. The expanding capacity of many large Chinese shipyards is likely to benefit the PLAN as its needs for more and larger ships grow or if it seeks to adapt merchant vessels for military operations. Beyond these technical issues, the profits generated by civilian production may also subsidize military production. Cross-subsidization has occurred in other defense industries such as aerospace and aviation.

China's shipbuilding industry also suffers from limitations and weaknesses that hamper naval modernization. Although the design and construction of vessels have improved, Chinese shipbuilders have experienced many problems producing quality subsystems for merchant and naval vessels. They have had to rely heavily on foreign imports for power plants, navigation and sensor suites, and key weapons systems for their newest naval platforms. For example, Chinese marine engine factories have had difficulties producing gas turbine engines powerful enough for large destroyers. The last two classes of Chinese destroyers have both relied on imported gas turbine engines. This high degree of reliance on foreign subsystems creates challenges for systems integration and complicates serial production of some platforms because of the potentially uncertain availability of certain subsystems.

The greatest weaknesses of China's naval platforms have been in their weapon systems. Chinese vessels have lacked long-range air defense systems, modern antisubmarine warfare weapons, and advanced electronic warfare capabilities. Those systems that are deployed tend to be copies or modifications of Soviet or Western systems. Few plants have been able to produce modern versions of these crucial systems. For example, Chinese suppliers have experienced repeated delays in the indigenous production of medium and long-range SAM systems for area defense; these delays in turn have delayed the completion of current naval projects.

In short, Chinese shipbuilders have been able to produce better-designed and better-fabricated warships in less time than previously, but these new platforms lack the advanced weapons, electronics, and propulsion subsystems needed to properly outfit these vessels. It is these technologies (and their integration) that will ultimately determine the PLAN's military efficacy. There are signs, however, that China is beginning to acquire the capability to produce some of these critical subsystems, especially with regard to air defense systems.

Information Technology

China's IT sector should be viewed as a civilian industry with links to the Chinese defense industrial establishment and the PLA. Certain IT companies supply finished command, control, communications, computers, and intelligence (C4I) equipment and related products to the PLA, facilitating a major modernization of China's military C4I infrastructure. Whereas China's defense-industrial system has long suffered from a wide-ranging set of structural problems that have impeded development of modern military equipment, the commercial IT sector carries none of these burdens. The industry is marked by new facilities in dynamic locales, a high-tech work force, and infusions of foreign technology. It does not have to worry about maintaining the social safety net for thousands of unemployable workers and their families in rural areas but instead attracts staff using market-based incentives, including stock options, and fires staff when necessary.

Using the siren song of China's IT market as a lure for acquiring cutting-edge foreign technology, China's IT companies, state R&D institutes, and state R&D funding have formed a potent "digital triangle" that combines the significant resources of the state with the market-driven dynamism of the commercial sector. The digital triangle is facilitated by a combination of a national development strategy focused on technology, high-level bureaucratic coordination, and significant fiscal support from national five-year plans and state science and technology budget programs like China' recent *863 Program*. Because of the fungibility of the technology, the IT sector is uniquely placed to exploit these trends by commercializing portions of the state R&D base for the benefit of both the civilian economy and military procurement. This civilianization of research involving military technologies is the real paradigm shift at the heart of the digital triangle: introducing commercial and profit-seeking motives as engines of innovation and increased efficiency to improve China's overall technological level and thereby indirectly benefit the military's IT capabilities.

The effectiveness of the digital triangle in achieving IT break-throughs can be seen in four areas: telecommunications equipment, supercomputers, core routers, and fiber optics. Foreign companies have aided the efforts of the digital triangle by infusing technology, equipment, capital, and know-how into important commercial companies linked to the system, mainly in an attempt to secure access to the Chinese market, because the nature of the regulatory and commercial environment in China places enormous pressure on foreign companies to transfer technology. Transfers of foreign technology have made the government ministries and companies less dependent on foreign money and technology over time. This dynamic is what Chinese interlocutors refer to as the "new model" or path to development for China: cooperation, learning, mastering, independent development, replacement, indigenous innovation, and, ultimately, global competitiveness.

The experience of procuring military-related IT equipment from Chinese companies has taught the General Armament Department a great deal about contracting, competition, and bidding, and has emboldened procurement officials to apply lessons from this sector to the traditional defense-industrial producers.[88] Yet Chinese officials are also quick to point to the limits or constraints of wholesale transfer of these lessons to the more traditional defense-industrial sectors in light of the unique characteristics and advantages of the IT sector over its far less nimble and dynamic counterparts, which are burdened with maintaining social stability at the expense of economic efficiency.

For the PLA, the digital triangle has opened a window of access to advanced information technologies, fueling a C4I revolution in the armed forces. As a result, the PLA has reportedly achieved significant improvements in its communications and operational security, as well as its capacity to transmit information, but it is not clear whether this increasingly advanced information technology system in the military will only improve the handling of information, or will perform the much larger function of boot-strapping the PLA's much more primitive, much less "informationized" conventional forces into a more modern force.

---

[88] Interviews with General Armament Department officials, Beijing, China, September 2000-September 2001.

Missile Industry

China's missile sector is generally regarded as one of the more successful and capable parts of the defense industry, a bright star in an otherwise mediocre economic sector. Many products from the missile sector are comparable in quality to those of Western nations. China's solid-fuel conventional ballistic missiles are increasingly reliable and accurate and have become a central element of some of the PLA's options in a Taiwan scenario. Similarly, China's *Ying Ji* (Eagle Strike) series of antiship cruise missiles are considered comparable in capability to the French Exocet or the U.S. Harpoon. China may also soon deploy its first land attack cruise missile. China is in the process of developing a number of new missile systems, including more-accurate and longer-range ballistic missiles, land-attack cruise missiles, and a long-range surface-to-air missile system comparable to the U.S. Patriot or Russian S-300 (SA-10/SA-20) series.[89]

Several factors explain the missile industry's relative success in improving the quality of its equipment and technologies over the past decades. Most of China's missiles are produced by subsidiaries of two large state-owned holding companies—China Aerospace Science and Technology Corporation (CASC) and China Aerospace Science and

---

[89] Duncan Lennox, "CSS-7 (DF-11/M-11)," *Jane's Strategic Weapon Systems 40*, 3 June 2003, (online at http://online.janes.com/, accessed 25 November 2003); Duncan Lennox , "CSS-6 (DF-15/M-9)," *Jane's Strategic Weapon Systems 40*, 3 June 2003, online at http://online.janes. com, accessed 25 November 2003; CSS-N-8 'Saccade' (YJ-2/C-802); CY-1/C-803)," *Jane's Naval Weapon Systems 39*, 28 August 2003, online at http://online.janes.com/, accessed 25 November 2003; Duncan Lennox, "CSS-N-4 'Sardine' (YJ-1/-12/-82 and C-801) and CSSC-8 'Saccade' (YJ-2/-21/-22/-83 and C-802/803)," *Jane's Strategic Weapon Systems* 40, 31 July 2003, online at http://online.janes.com/, accessed 25 November 2003); Duncan Lennox, "YJ-1 (C-801) and YJ-2 (C-802)," *Jane's Air-Launched Weapons* 38, 09 November 2001, online at http://online.janes.com/, accessed 25 November 2003; Duncan Lennox, "MM 38/40, AM 39 and SM 39 Exocet," *Jane's Strategic Weapon Systems 40*, 31 July 2003, online at http://online. janes.com/, accessed 30 November 2003; Duncan Lennox, "AGM/RGM/UGM-84 Harpoon/ SLAM/SLAM-ER," *Jane's Strategic Weapon Systems 40*, 27 October 2003, online at http://online.janes.com/, accessed 30 November 2003; Rob Hewson, "YJ-6/C-601 (CAS-1 'Kraken')," *Jane's Air-Launched Weapons* 40, 9 July 2002, online at http://online.janes.com/, accessed 25 November 2003; Duncan Lennox, "HQ-9/-15, HHQ-9A, RF-9," *Jane's Strategic Weapon Systems 39*, 6 January 2003, online at http://online.janes.com/, accessed 25 November 2003; James C. O'Halloran, "Chinese self-propelled surface-to-air missile system programmes," *Jane's Land-Based Air Defence*, 27 January 2003, online at http://online.janes.com/, accessed 25 November 2003; U.S. Department of Defense, 2004.

Industry Corporation (CASIC).[90] These companies are organized differently from most of China's other defense conglomerates. Rather than simply aggregating dozens of research institutes, enterprises, factories, and other subordinate companies, CASC and CASIC each contain 6–8 "research academies" and "bases." These academies and bases have autonomous financial and auditing powers, management rights, and responsibility for profits and losses. Within each academy or base are numerous subordinate entities including research institutes, factories, schools, hospitals, and subsidiary companies (research institutes tend to predominate in the academies, whereas factories tend to play a more important role in the bases, but all academies and bases contain both research institutes and factories).

This organizational structure is in part the legacy of the key role the missile sector played in China's strategic weapons programs. China's missile sector has consistently received funding and significant political support over the years. More recently, China's missile producers have benefited from the large appropriations for China's manned space program. Finally, China's missile manufacturers have been able to generate revenues through exports of missile systems and the sale of satellite launch services to foreign buyers.

Despite these advantages, China's missile sector suffers from some of the same problems as China's other defense industry sectors. The only indigenously produced long-range surface-to-air missiles in service in China are improved versions of the 1950s-era Soviet SA-2 system. Similarly, until recently, China did not have a solid-fuel intercontinental ballistic missile (a capability that the United States has had since the early 1960s), land-attack cruise missiles, beyond-visual-range air-to-air missiles, or antiradiation missiles.

These capability shortfalls stem from many of the same weaknesses that exist in China's other defense industries, including weak

---

[90] In addition to missiles, CASC and CASIC produce systems for China's space program and are often referred to as China's "aerospace" industry by English-language publications of the Chinese government and translations of Chinese writings on the topic. However, these firms do not produce aviation products, so the translation is inconsistent with the English meaning of the term "aerospace." Moreover, the Chinese term used to characterize this sector—航天 (literally, "to navigate the heavens")—is more usually translated as "spaceflight" or "astronautics." As in China's other defense sectors, CASC and CASIC were created in 1999 by the bifurcation of a predecessor, in this case a company called the China Aerospace Corporation.

linkages between research and production and minimal incentives for innovation and efficiency. CASC and CASIC are also inefficient, and their workforces are bloated. They have been adopting policies to improve their performance, however, such as attempts to increase the accountability of enterprise managers; strengthen financial controls; and attract, retain, and promote talented personnel. Enterprises are also encouraged to engage in the production of civilian goods and to cooperate with foreign firms, but the nature of the industry is likely to limit international cooperation. Unlike the aviation industry, Western firms have shown little interest in subcontracting work to China's space sector, in great part because U.S. export controls severely constrain the types of technology U.S. firms can share with Chinese counterparts.

Like China's other defense sectors, China's missile industry is showing signs of rapid improvement. Many long-standing shortcomings are expected to be rectified in the next few years, at least partially. By the end of this decade, China's military is expected to field indigenously produced mobile, solid-fuel ballistic missiles; land-attack cruise missiles; modern, long-range surface-to-air missiles; beyond-visual-range air-to-air missiles; and antiradiation missiles. This array of systems, if successfully deployed, would enable China to equip its military with missile capabilities comparable to those of all but the most advanced militaries in the world.

Future Prospects of China's Defense Industry

Over the past two decades, China's defense industries have made substantial progress in improving the efficiency of their operations and the quality and technological sophistication of their products. As measured by improvements in the technological quality of the output of these industries, this process has accelerated over the past five years. These improvements suggest a defense industry that is beginning to emerge from the doldrums of two decades of stagnation and neglect. This progress has been facilitated in part by organizational changes that have created incentives for managers to improve efficiency and accelerate innovation. Other key factors include increased government expenditures on research, development, and procurement; increased imports of foreign equipment, capital and know-how; and the "spin-on" effects of commercial business operations. In particular, those industries with robust and rational commercial activities, especially those linked to international markets, have shown the greatest improvements in capabilities.

China's emerging IT sector, although not officially part of China's defense industrial complex, is the most organizationally innovative and economically dynamic of China's producers of military equipment. While the industry is primarily oriented toward China's commercial market, the PLA has been able to effectively leverage the industry's production capabilities to facilitate improvements in the military's C4I capabilities—a critical element of the PLA's modernization efforts. The IT sector will continue to follow world standards in telecommunications equipment, supercomputers, routers, and other technologies; the Chinese military will have access to these capabilities for critical C4I applications.

Although China's shipbuilding industry has been burdened with many of the trappings of a militarized planned economy, it has rapidly expanded exports and as a consequence has gained increasing access to foreign technologies, capital, and know-how. As its commercial business has expanded, naval production has also benefited. China's shipbuilding industry now produces a wide range of increasingly sophisticated naval platforms using modern design methods, production techniques, and management practices that were developed for the construction of civilian ships. These improvements are likely to continue in the coming years. However, the shipbuilding industry still suffers from some major failings. It still lacks the ability to build critical naval subsystems, limiting the overall warfighting capabilities of naval vessels produced in China.

Technological progress in China's missile industry appears to have accelerated in the past five years. The missile sector continues to produce new and increasingly advanced ballistic and antiship cruise missiles. It will soon begin fielding land attack cruise missiles, modern, long-range surface-to-air missiles, beyond-visual-range air-to-air missiles, and antiradiation missiles.[91]

---

[91] Duncan Lennox, "CSS-7 (DF-11/M-11)," *Jane's Strategic Weapon Systems 40,* 3 June 2003, online at http://online.janes.com/, accessed 25 November 2003; Duncan Lennox , "CSS-6 (DF-15/M-9)," *Jane's Strategic Weapon Systems 40,* 3 June 2003, online at http://online.janes.

In recent years, signs of progress have begun to emerge in China's aviation industry. China's first indigenously designed combat aircraft has recently entered service. China is on the verge of producing an indigenously developed fourth generation aircraft, albeit with substantial foreign assistance. It is also expected to begin producing its first operational turbofan engines, possibly ending its dependence on imported engines to power its more modern combat aircraft. Important gaps in China's aviation design and production capabilities remain, however. China is still unable to produce heavy bombers or large transport aircraft. It has yet to field an indigenously designed helicopter. Important structural weaknesses in China's aviation industry still inhibit advances in R&D. Although the industry may be narrowing the gap with the world's most advanced nations, it shows no signs of achieving parity with those industries for the foreseeable future.[92]

---

com, accessed 25 November 2003; CSS-N-8 'Saccade' (YJ-2/C-802); CY-1/C-803)," *Jane's Naval Weapon Systems 39,* 28 August 2003, online at http://online.janes.com/, accessed 25 November 2003; Duncan Lennox, "CSS-N-4 'Sardine' (YJ-1/-12/-82 and C-801) and CSSC-8 'Saccade' (YJ-2/-21/-22/-83 and C-802/803)," *Jane's Strategic Weapon Systems* 40, 31 July 2003, online at http://online.janes.com/, accessed 25 November 2003; Duncan Lennox, "YJ-1 (C-801) and YJ-2 (C-802)," *Jane's Air-Launched Weapons* 38, 09 November 2001, online at http://online.janes.com/, accessed 25 November 2003; Duncan Lennox, "MM 38/40, AM 39 and SM 39 Exocet," *Jane's Strategic Weapon Systems 40,* 31 July 2003, online at http://online.janes.com/, accessed 30 November 2003; Duncan Lennox, "AGM/RGM/UGM-84 Harpoon/SLAM/SLAM-ER," *Jane's Strategic Weapon Systems 40,* 27 October 2003, online at http://online.janes.com/, accessed 30 November 2003; Rob Hewson, "YJ-6/C-601 (CAS-1 'Kraken')," *Jane's Air-Launched Weapons* 40, 9 July 2002, online at http://online.janes.com/, accessed 25 November 2003; Duncan Lennox, "HQ-9/-15, HHQ-9A, RF-9," *Jane's Strategic Weapon Systems 39,* 6 January 2003 online at http://online.janes.com/, accessed 25 November 2003; James C. O'Halloran, "Chinese self-propelled surface-to-air missile system programmes," *Jane's Land-Based Air Defence,* 27 January 2003, online at http://online.janes.com/, accessed 25 November 2003; U.S. Department of Defense, May 2004.

[92] Munson, "HAI (Eurocopter) Z-9 Haitun," op. cit.; Wang, "New Military Aircraft Displayed at the National-Day Grand Military Parade," *Liaowang,* 8 November 1999, pp. 30–31; Xu Dashan, *China Daily* (internet version), 13 September 2001, in FBIS as "China Daily: Helicopter Sector To Be Promoted," 13 September 2001; Xu Dashan, *China Daily* (internet version) 11 July 2002, in FBIS as "PRC's New H410A Helicopter Model Receives Certification; CAAC Says 'Huge Achievement,'" 11 July 2002; Robert Sae-Liu, "China advances helicopter projects," *Jane's Defence Weekly,* 3 May 2002; Munson, "CHAIG Z-11," op. cit.; Munson, "CHAIG Z-8," op. cit.; Munson, "CHRDI Z-10," op. cit.; Munson, "CAC J-10," op. cit.; "Chinese Puzzle," *Jane's Defense Weekly,* op. cit.; Yihong Chang, "China launches new stealth fighter project," *Jane's Defence Weekly,* 11 December 2002; U.S. Department of Defense, 2004.

Over the past few decades, China has trained and developed a cadre of highly capable scientists, engineers, and technicians. It has also invested in modern manufacturing facilities in many key plants in the defense sectors. These capabilities, which take years or even decades to build, are now in many respects competitive with those of Western countries. However, the organizational, contracting, incentive, and other systems in China are such that these relatively high-quality inputs still do not generate weapon systems of a comparable degree of sophistication and capability. In our view, although these institutional shortcomings remain serious, further institutional reforms could serve to rectify the problems. Given the proper incentives, China's defense industries could become much more competitive. A number of indicators suggest that key defense sectors are already overcoming long-standing weaknesses.

As noted in Chapter Four, there is considerable budgetary evidence that the Chinese government has been rapidly increasing military expenditures, including expenditures on procurement. Some of these expenditures appear to have been used to improve the quality of physical capital in the industry. Similarly, by breaking defense industry corporations into semiautonomous enterprises able to compete and participate in open bidding for contracts, China is on the verge of introducing true competition into the defense sector. The Chinese government is also beginning to encourage enterprises to improve the quality of their labor forces by granting permission to freely hire and fire employees, although this transformation will take time. China's civilian economy possesses a large and growing pool of technical talent that could be convinced to work for the defense sector if offered attractive levels of compensation.

Some of these changes have already taken place, triggered by the most recent round of reforms initiated in 1998. The effects are beginning to be reflected in the operations and output of key sectors in China's defense industry. If these trends persist, China's defense industry should be able to increasingly provide China's military with more capable, more technologically sophisticated weapons and equipment.

CHAPTER SIX

PLA Threat Perceptions and Force Planning

Up to this point in our analysis, we have focused on factors that would make it possible for Chinese leaders to accelerate the modernization of their military. We projected likely rates of growth in the Chinese economy and assessed pressures for increased government spending on areas other than the military. We then evaluated current spending on the military, examining budgeting processes and assessing sources of revenues for military spending outside the official military budget. We concluded by providing our own estimates of the current size of military spending.

We then turned to an assessment of what these funds are able to buy. We identified key weaknesses in the procurement process and contracting procedures but also identified institutional changes that already appear to be leading to improvements in the quality and capabilities of domestically produced weaponry. We then conducted a detailed evaluation of four key military industries and found that overall quality is improving but that, outside the IT sector, which is not formally a sector within the defense industries, China still has difficulty in producing highly capable military systems and weapons.

In this chapter, we examine the question of demand for military capability. Future defense spending in China will not be driven just by the size of the economy and expanding government budgets. There are too many politically powerful claimants on government resources. Chinese government and PLA perceptions of threats and their desire to use military capability to influence regional political relations will have a major effect on decisions about future military spending. Although we cannot translate government or PLA desires into precise forecasts of future military expenditures, we can use analysis of Chinese threat perceptions to provide indications of the relative ranking of military spending in the hierarchy of demands placed on the Chinese government.

We first analyze the threat perceptions upon which Chinese military leaders base their plans for current and future force structures. We then describe the hierarchy of immediate and potential security concerns that inform PLA planning and force structure modernization. Based on this analysis, we link these threat perceptions to general trends in Chinese defense procurement. We conclude by identifying key procurement areas that the Chinese military is currently emphasizing to meet perceived threats and challenges to national security.

PLA Threat Perceptions and the International Security Environment[1]

An assessment of PLA threat perceptions begins by understanding the PLA's conception of Chinese national security and national interests.[2] Chinese military strategists consistently emphasize the need to maintain the existence of three "conditions" for China to survive and prosper. In order of importance, the three conditions are: national unity, stability, and sovereignty. PLA threat perceptions and strategic planning are broadly informed by the need to maintain these three conditions. The PLA's assessment of the international security environment is based on the extent to which the policies and actions of other nations directly or indirectly challenge or threaten China's ability to maintain unity, stability, and sovereignty.

---

[1] This section draws on analysis from U.S. Department of Defense, 2003; Council on Foreign Relations, 2003; and Shambaugh, 2003, especially Chapter 7.

[2] The Chinese military's views of the international security environment and its threat perceptions should not necessarily be seen as synonymous with and identical to the views of China's senior political leaders. Given that many of China's senior political leaders lack military experience and view the PLA as one of many constituencies in China's national security bureaucracy, they may differ with the PLA concerning the scope and intensity of certain threats facing China. The point of this section is to highlight the views of PLA leaders to understand the specific threat assessments that are driving defense procurement and military modernization.

These themes were explicitly and more fully addressed for the first time in China's most recent *National Defense White Paper*.[3] This document, which was last published in December 2002 and is produced evey two years, outlined a set of national interests that serve as "the fundamental basis for the formulation of China's national defense policy."[4] These include safeguarding state sovereignty, unity, territorial integrity, and security; upholding economic development as the central task and unremittingly enhancing the overall national strength; adhering to and improving the socialist system; maintaining and promoting social stability and harmony; and striving for an international environment of lasting peace and a favorable climate on China's periphery. PLA perceptions of current and future threats are primarily based on this collection of national interests.[5]

Beyond these broad themes, the writings of Chinese military officers and official government assessments suggest a range of specific threats and potential challenges to Chinese security.[6] These perceptions drive current and future directions in doctrine and force structure planning. The most important threats for the PLA currently include

- U.S. military and foreign policies (especially those related to Taiwan)
- Japan's reemergence as a regional military power
- India's growing military power and regional influence
- border and coastal defense
- defending territorial waters and airspace.

---

[3] Curiously, China's biannual defense white paper is published not by the PLA but rather by the State Council's Information office. This strongly suggests that the views expressed in this document have been vetted by senior civilian leaders as well as military ones.

[4] *China's National Defense in 2002*, p. 11; Wang, 1999.

[5] *China's National Defense in 2002*, p. 11.

[6] Useful collections of PLA assessments include: 2000–2001 *Zhanlue Pinggu* [2000–2001 Strategic Assessment], Beijing, China: Junshi Kexue Chubanshe, 2000; *Yatai Anquan Zhanlue Lun*, Beijing, China: Junshi Kexue Chubanshe, 2000; Michael Pillsbury, *Chinese Views of Future Warfare*, Washington, D.C.: National Defense University Press, 1998; Michael Pillsbury, *China Debates the Future Security Environment*, Washington, D.C.: National Defense University Press, 2000.

First and foremost, PLA military strategists perceive the United States as posing both an immediate and long-term challenge to Chinese national security interests. This perception is based on a set of concerns about U.S. policies on the Taiwan issue (considered most important), U.S. alliance relationships and defense ties in Asia, and overall U.S. national security strategy. Although China's publicly articulated concerns about the United States have subsided since 2001, there are a number of indications that the PLA and the Chinese leadership continue to view the United States as the major challenge to Chinese national security.

The Chinese leadership's and the PLA's fears about Taiwanese independence and possible U.S. intervention are far and away the most immediately relevant to the PLA's current planning and procurement.[7] Since the end of the 1990s, PLA reform, modernization, procurement, and training have been heavily—almost singularly—focused on preparing for a conflict over Taiwan. Concerns about a conflict over Taiwan's status are acute and in a category of their own for the military: Taiwan is at the top of the PLA's watch-list of possible conflicts. In this context, U.S. policies on the Taiwan question are of immediate concern to Chinese defense planners. The leaders of the People's Republic of China are committed to ensuring the reunification of Taiwan on their terms; U.S. policies are seen as directly preventing this outcome. Some Chinese military planners fear that the United States seeks to keep Taiwan apart from the mainland to use it as a strategic point in Asia to limit the growth of China's regional influence. Specifically, PLA strategists perceive U.S. arms sales to Taiwan and bilateral military interactions as part of a determined effort to keep China permanently divided and thus undermine key Chinese goals of unity and sovereignty. If a conflict were to erupt over Taiwan, U.S. intervention is a particularly worrisome possibility for PLA strategists. Most Chinese and Western analysts presume that the United States would intervene in a conflict, unless Taiwan declared independence without provocation from the mainland. As a result, much of the PLA's modernization has been focused not only on fighting Taiwanese forces but also fighting U.S. forces if a conflict were to erupt.

---

[7] See Department of Defense, 2003, pp. 1–16.

For example, in recent years the PLA has been developing "asymmetric capabilities" to deter or degrade superior U.S. military capabilities in the event of the outbreak of a conflict.

Beyond the narrow Taiwan contingency, Chinese military planners and political leaders are decidedly uncomfortable with the U.S. military presence in the world. They fear that the United States can and will use military force whenever and wherever it wants, including in scenarios involving Chinese security interests. According to the 2002 U.S. Defense Department report on the PLA, "China's leaders have asserted that the United States seeks to maintain a dominant geo-strategic position by containing the growth of Chinese power, ultimately "dividing" and "Westernizing" China, and preventing a resurgence of Russian power."[8] These concerns are often articulated in the Chinese "code language" of "the rise of hegemony and power politics" in international relations. Chinese leaders often argue, with varying degrees of fervency based on the overall climate in U.S.-China relations, that "hegemony and power politics" is on the rise and that China and other nations need to challenge these trends. Chinese concerns about U.S. unipolarity and its military policies were most acute in 1999–2001, especially following the accidental NATO bombing of China's embassy in Belgrade. Since mid-2001, the number and tone of Chinese public statements about U.S. intentions have moderated substantially.[9]

PLA concerns about U.S. policies are also often raised in the context of U.S.-Japanese relations and to a lesser extent U.S. defense ties with Southeast Asian nations. Regarding these countries, Chinese strategists are most concerned about U.S. alliances, U.S. military aid, U.S. missile defense cooperation, and the overall constellation of U.S. defense relationships in the region. PLA planners see U.S. military ties with Asian nations as part of a U.S. strategy to maintain a dominant and controlling strategic position in Asia in order to limit the growth of China's military power and economic influence. Specifically, Chinese military planners interpret the efforts of the United States to bolster its alliance relations with Japan and Southeast Asian nations as a manifes-

---

[8] Department of Defense, 2003, p. 8.

[9] This claim is based on the author's comparisons of the 2000 and 2002 Chinese Defense White Papers.

tation of a broader U.S. strategy to contain China's power and influence in the region.

A second and related security concern for Chinese military planners is Japan. Although Chinese political leaders continue to value Sino-Japanese economic relations for their contribution to domestic growth, Chinese military strategists remain concerned about the possible rebirth of Japanese militarism and about Japan's military alliance with the United States. In light of the long, violent and sordid Japanese occupation of Chinese territory in the first half of the twentieth century, Chinese leaders and military strategists are acutely concerned about the reemergence of Japan as a military power. They see changes in Japanese military doctrine, force structure, and deployment in recent years as evidence of Japanese efforts to improve Japan's military capabilities and assume a more influential role in Asia. According to one authoritative study of PLA threat perceptions, "The anti-Japanese sentiment one encounters among the PLA at all levels is palpable. Distrust of Japan runs deep, transcends generations, and is virulent among the generation of PLA officers in their fifties and sixties."[10] U.S.-Japan military relations are also interpreted in that context. PLA strategists fear that the United States may unwittingly facilitate Japanese rearmament through bilateral trade in weapons and defense technologies, especially through cooperation on missile defense. An equally acute fear on Beijing's part is that the U.S.-Japanese military alliance is a central part of a regional strategy to limit China's influence in Asia and to contain the "China threat."

India is a third and relatively new security concern for Chinese military planners. China and India fought a border war in 1962. Since then the PLA has deployed troops along the Sino-Indian border to forestall India from attempting to regain lost territories. Yet, for the past several decades, the border has been relatively stable and India has not posed a major security threat to China. As a result, PLA specialists have written relatively little about India, especially compared with the numerous books and articles about the United States, Taiwan, and Japan.

---

[10] Shambaugh, 2003, p. 301.

India's nuclear tests in 1998 rang an alarm bell of sorts for the PLA. The fact that the Indian Defense Minister justified these tests as a response to the growing threat posed by China aggravated the perception of PLA strategists that India was becoming a threat. Since 1998, Chinese military writings have reflected a growing concern about India's nuclear and missile capabilities. They question whether New Delhi will seek to use its growing nuclear capabilities to coerce China. An emerging concern in Chinese military circles is the U.S.-Indian military-to-military relationship.[11] Many in the PLA have begun to view U.S.-India military relations as increasingly directed at containing China. Interestingly, the PLA's heightened concerns are emerging at the same time as Chinese leaders are making a determined effort to improve Sino-Indian political relations.

Defending China's borders and coastlines and protecting China's territorial claims in the South China Sea constitute the fourth and fifth categories of potential missions and concerns for PLA strategists. China shares land borders with fourteen countries and maritime boundaries with seven. China's borders have long been a source of significant concern for its military leaders: China has fought significant border-related conflicts with India, Russia, and Vietnam and has had low-intensity territorial disputes with at least ten other nations.[12] Throughout the 1990s, China actively pursued numerous territorial disputes that have historically been the cause of substantial regional tension. However, since 1991, China has settled disputes with Laos, Russia, Kazakhstan, Kyrgyzstan, Vietnam, and Tajikistan. China has also been actively involved in the establishment and expansion of a regional security organization among China, Russia, and Central Asian nations known as that Shanghai Cooperation Organization. The organization was set up to address mutual concerns about border security and terrorism in the region.

Protecting Chinese territorial waters and airspace has long been a primary mission for the Chinese military. This mandate emerged as

---

[11] Yuan, 2001, pp. 978–1001; see also Sighu and Yuan, 2003.

[12] These include: Nepal, Sikkim, Burma (Myanmar), Thailand, Malaysia, Indonesia, the Philippines, Brunei, Laos, Kazakhstan, and Kyrgyzstan.

paramount for the PLA at the very founding of the PRC in 1949 when Nationalists troops launched incursions into the mainland from Taiwan. The 2001 EP-3 incident between the United States and China heightened the PLA's focus on defending Chinese territorial waters and its airspace. China's heavy emphasis on improving its air defense capabilities in recent years is a reflection of its growing concern about defending its borders, both land and sea.

Another PLA mission related to protecting Chinese sovereignty is defending Chinese territorial claims in the South China Sea. Beginning in 1988, China began to systematically reassert its claims to areas in the South China Sea. China now occupies seven "features": rocks, reefs, shoals, and islands in the South China Sea. China has made claims to other territories in this sea. PLA literature on defending Chinese sovereignty often mentions China's territorial claims in the South China Sea as one of its areas of responsibility.[13]

Analyzing and Interpreting PLA Threat Perceptions

Several important themes emerge from this assessment of PLA threat perceptions. Many of them are not readily apparent when assessing the various nations and contingences that the PLA views as immediate threats or potential challenges to national security. First, compared to the last several decades, China has never been more secure. China has resolved many of its most volatile border disputes (e.g., with Russia and Vietnam) and has made great progress in forging stable and cooperative relations with most neighboring counties. It has become involved and active in several Asian regional security organizations in an effort to build confidence about security issues. The dramatic improvement in Sino-Russian relations leading to the establishment of a "strategic cooperative relationship" in 2001 is most notable in this regard. These realities stand in stark contrast to China's systemic insecurity in the past. Throughout the first half of the twentieth century China, was invaded or occupied by various Western and Asian nations.

---

[13] Some of these scenarios are outlined in Wang Hongqing and Zhang Xingye, eds., *Zhanyixue* [The Science of Military Campaigns], Beijing, China: Guofang Daxue Chubanshe, May 2000; also see Shambaugh, op. cit.

Just after the People's Republic of China was founded, it entered into a bloody conflict in Korea. During the 1960s and 1970s, China and the Soviet Union stood on the brink of full-scale, possibly nuclear, war several times. Beijing's current security situation differs starkly from past years.

Second, major Chinese public assessments of the international security environment can change quickly and thus are not necessarily an entirely reliable indicator of the long-term perceptions of Chinese military planners. In particular, the tone of Chinese assessments of the international security environment vacillates with the overall political climate in U.S.-China relations. The difference in the tone of the 2000 and 2002 National Defense White Papers is one such example. The language in the 2000 Defense White Paper was strident and acerbic in its characterization of the threats and challenges that the United States posed for Chinese and world security. The language was clear and explicit about U.S. policies serving as the main source of instability and insecurity in international relations. By contrast, just two years later in 2002 and a year after the 9/11 incident, the 2002 Defense White Paper was far more moderate about the international security environment. The 2002 version carried much less sharp characterizations of threats facing China, downplayed the threats posed by "hegemony and power politics," seldom referred to the United States by name, and emphasized the importance of great power cooperation as a new and growing trend in global politics. Thus, official public documents outlining threat perceptions should be seen as political documents. Their tone is more a reflection of the state of China's relationship with other major powers, such as the United States, Russia, and Japan, and less an indication of a wholesale shift in the PLA's worldviews—let alone those of China's political leaders.

Third, in recent years, PLA threat assessments have reflected a new recognition of the challenges posed by nontraditional security issues such as terrorism, arms control, the proliferation of weapons of mass destruction, drug-trafficking, and environmental issues, among others. PLA writings have indicated an awareness that transnational topics could be sources of instability that threaten China's national security interests. These are new themes in PLA writings, which in the past focused on the traditional security threats posed by other nations. This development is in part an outgrowth of China's emphasis on ensuring "comprehensive security," which links national security to China's overall economic development and the government's ability to build a "well-off society." PLA conceptions of national security and national interest appear to be broadening. However, a common theme in PLA assessments of transnational security challenges is that U.S. policies exacerbate many of them.[14]

PLA Force Structure and Military Modernization[15]

The PLA seeks to modernize its force structure to provide it with the capabilities to meet the threats and challenges noted above. The PLA seeks four categories of capabilities:

1. The capability to respond to both internal and external threats by quickly taking the initiative, preventing escalation, attaining superiority, and resolving the conflict on China's terms
2. The eventual development of a limited power projection capability that would facilitate a sustained sea presence and an area denial capability, although area control is not a high priority for the PLA
3. The ability to conduct short-range preemptive strikes using conventional missiles and air force assets
4. The development of a credible strategic nuclear capability to deter other nuclear powers from using nuclear threats to coerce China or to limit its strategic options, especially during a crisis.

To carry out missions, PLA leaders have sought to develop a more modern, diverse, versatile set of military capabilities. The PLA's desired force structure consists of two main components: conventional and strategic nuclear forces. The difference between actual and desired

---

[14] Shambaugh, 2003, p. 297.

[15] This section is largely drawn from the analysis and conclusions of U.S. Department of Defense, 2002, and Council on Foreign Relations, 2003. Both reports outline the force structure needs of the PLA and the priorities of the PLA in its modernization efforts.

capabilities should also be viewed as current gaps or weaknesses in the PLA's ability to carry out its core missions.

Regarding conventional weapons capabilities, the PLA is currently transitioning from a continental military requiring large land forces for "in-depth" defense to a combined continental-maritime force primarily consisting of smaller, more mobile and sophisticated military forces. To address the immediate threat posed by Taiwan, the PLA specifically needs to acquire significant disruption and strike capabilities against the island including area denial capabilities against U.S. forces to complicate U.S. efforts to deploy forces in the vicinity of Taiwan. These are often termed "anti-access" capabilities. Thus, the PLA needs to develop or modernize a host of specific conventional capabilities (1) to facilitate the shift to a continental-maritime force over the medium to long term and (2) to develop a capability to address a Taiwan contingency in the immediate future. The PLA seeks to acquire the following specific conventional capabilities:

- A smaller, more flexible, highly trained and well-equipped ground force centered on rapid reaction units, with greater numbers of airborne troops and substantially larger amphibious lift capabilities.
- A robust green-to-blue-water naval capability centered on a new generation of surface combatants with improved air defense, antisubmarine warfare, and antiship capabilities; modern, quieter conventional attack submarines with advanced torpedoes and cruise missile capabilities; significantly enhanced replenishment-at-sea capabilities; and larger and more modern naval air assets.
- A more modern and versatile air force that possesses more advanced longer-range strike and ground-attack aircraft, improved early warning and air defenses, extended and close air support, and longer range transport, lift, and mid-air refueling capabilities.
- A joint service tactical operations doctrine utilizing more sophisticated C4ISR, early warning, and battle management systems, and both airborne and space-based assets, to improve detection, tracking, targeting, and strike capabilities and to enhance operational coordination among the armed services.

China's development and deployment of significant numbers of conventionally armed short- and medium-range ballistic and cruise missiles is directly linked to Chinese force planning for a Taiwan conflict. Such systems offer China its most potent form of coercive capability against the Taiwanese government and the Taiwanese people. Other desired conventional capabilities of direct relevance to the Taiwan situation include the following:

- Longer-range antiship cruise missiles (ASCMs) and land-attack cruise missiles (LACMs)
- Improved information operations and electronic warfare capabilities
- Fourth generation strike fighters
- Advanced diesel submarines
- Advanced multirole destroyers with ASCMs
- Improved communication capabilities
- Limited anti-satellite capabilities.

In terms of strategic nuclear capabilities, China is transitioning from possessing a small, unsophisticated, and highly vulnerable nuclear force to a more modern force that is far more survivable and reliable. This is being accomplished through the development and deployment of new types of road-mobile, solid-fuel missiles that are far more reliable and accurate and have shorter readiness times than the previous generation. Some in China may also be contemplating the shift to a "limited nuclear deterrent" capability that would allow China to target military sites as part of a damage limitation strategy—as opposed to a nuclear strategy that simply seeks to provide a secure second-strike capability. To achieve a more credible nuclear deterrent the PLA needs to acquire the following capabilities:

- A greater number of land- and sea-based longer-range ballistic missiles with improved range, accuracy, and survivability to bolster the credibility of China's nuclear deterrent. The exact number and configuration of such systems will depend greatly on the structure and size of any future U.S. missile defense system

- More advanced warhead technologies that could penetrate a limited U.S. missile defense system
- Smaller, more powerful nuclear warheads with potential multiple independently targetable reentry vehicle (MIRV) or multiple reentry vehicle (MRV) capabilities. In the past, China eschewed developing this capability. Future decisions will be influenced by U.S. ballistic missile defense programs
- A modern early warning system with advanced land, airborne, and space-based C4ISR assets.

Collectively, the above conventional and nuclear capabilities represent the central areas of focus for China's ongoing military modernization effort. To be sure, certain projects have a much higher priority than others, such as the need to increase the numbers and accuracy of short- and medium-range ballistic missiles and to improve its C4ISR capabilities to be applicable to all possible military contingencies. The breadth of the PLA's desired capabilities raises questions about how and when the military will eventually be able to develop and deploy them. While the PLA has purchased many niche capabilities from Russia in recent years, indigenous procurement remains the preferred, long-term channel for the PLA to acquire more modern conventional and nuclear capabilities.

CHAPTER SEVEN

Future Expenditures on the Military

Introduction

The purpose of this chapter is to bound the likely future levels of resources that the Chinese government will provide its military. The absolute size of future Chinese government expenditures will be dictated by the size of China's economy, the ability of government to extract resources from it, and the government's decisions on how to allocate those resources. In Chapter Two, we forecast the future size of China's economy in terms of constant 2001 renminbi and then converted those projections into constant 2001 dollars using a combination of projected purchasing power parity and market renminbi/dollar exchange rates. In this chapter, we use these projections to estimate the future resource base on which military expenditure decisions will be based.

To project future military expenditures, we draw on the analysis in Chapter Three to assess future fiscal demands on the Chinese government. We focus on expenditures on those government services and transfer payments for which the Chinese government is most likely to face the greatest popular pressure for increases in the coming decades. We have selected what we see as among the most important of these items and projected potential future growth in government expenditures in these categories. The selected categories include health care, education, pensions, and the future interest that the Chinese government will have to pay on bonds issued to cover currently unfunded liabilities stemming from the financial problems in China's state-owned sector. We also project likely aggregate growth on other government expenditure categories, excluding military spending.

As noted in Chapter Six, government expenditures on the military are driven by military assessments of the capabilities needed to counter perceived threats and to attain national strategic objectives, as well as by resource constraints. We use the analysis in Chapter Six, coupled with our projections of future nonmilitary demands on the budget, to project upper-bound and mid-level forecasts of future military spending in constant renminbi and constant 2001 dollars through 2025.

We conclude the chapter by comparing projected future expenditure levels with past U.S. and USAF expenditures to provide a sense of what future Chinese expenditures are likely to purchase in aggregate and in terms of procurement of military equipment and weapons. This analysis is based in part on our assessment of the future purchasing power of Chinese budgetary expenditures for procurement. We argue that in the coming decades that the PLA will be able to obtain major weapons systems from China's domestic defense industry that will be of a quality and capability and at a cost comparable to those that the PLA currently imports from foreign suppliers such as Russia and Israel. We base this argument on our assessment of recent and likely impending improvements in the performance of China's defense industries presented in Chapter Five.

The Future Composition of Government Expenditures

China is becoming older, wealthier, and more urbanized. These changes are generating pressures for additional government spending in a number of categories: health care, pensions, public infrastructure, and the environment. The Chinese leadership's desire for rapid economic growth has led to a higher priority for expenditures on education as well. Simultaneously, economic reform and the ensuing displacement of workers and uneven distribution of incomes are generating pressures for expenditures on unemployment insurance and other government support for displaced workers and the redistribution of tax revenues from wealthier provinces to those that are less well-off. Finally, as noted in Chapter Three, the Chinese government faces a host of unfunded liabilities stemming from problems in the financial system that must be recognized in the near future. The government will have to service the increased debt that will be a consequence of these obligations. Future increases in military spending will have to vie with these needs for government resources. Below, we quantify some of these competing claims on government expenditures.

Future Expenditures on Education and Health
Government expenditures on education and health, along with science and culture, have risen along with economic output. However, their share of GDP has ebbed and flowed over the course of the past 25 years. In 1978, they ran 4.1 percent of GDP, then fell to 3.0 percent in 1995 and have since risen to 5.4 percent in 2001. The relative decline in public provision of education and health in the mid-1990s was driven by two trends. First, state-owned enterprises have been shedding responsibility for health care costs, forcing families to pay a far greater share of these costs privately. Second, the central government devolved more responsibility to municipalities for financing education and public health care after the tax reform in 1994. Many municipalities, especially those in poorer areas, found it difficult to fully fund either. In the case of education, municipalities imposed tuition fees and parents must pay for books. In some instances, destitute parents have chosen to pull their children from school at an early age because they could not afford tuition.

The Chinese leadership concluded that low levels of education, especially in poorer rural areas, had become a national scandal and a worry for leaders concerned about social stability. Consequently, they have begun to allocate more public money for education. Finance Minister Xiang's March 2003 Budget Report states that national expenditures for education increased 19.6 percent in 2002 and that there is a national policy of increasing the share of education expenditures in the central government's budget by one percentage point per year.

This policy has scored some successes. For the five-year period ending in 2002, education expenditures rose by 230 percent, an average annual rate of increase of 17.7 percent, according to the Finance Minister's 2003 Budget Report. China is successfully creating world-class universities in Beijing and Shanghai and obtaining world-class

Figure 7.1
Secondary School Enrollment

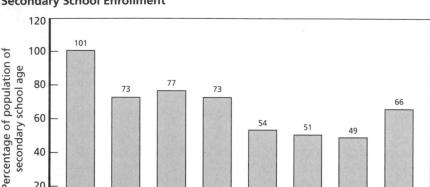

SOURCE: World Bank, World Development Indicators 2002, CD-ROM version.
RAND *MG260-7.1*

educations for much of its elite by sending students abroad. However, its performance in increasing enrollment rates at the high school and university levels is less impressive and hampers its aspirations to become a wealthy society. Although China does a better job at education in terms of secondary school enrollment rates than India and Thailand, its performance lags compared to that of Hong Kong, Singapore, South Korea, and Taiwan (Figure 7.1). If one compares tertiary school enrollment rates, China's performance is even poorer (Figure 7.2).

If China aspires to sustained economic growth and regional leadership, it will require massive catch-up investments in secondary and tertiary education. As Thailand has discovered, rapid economic growth can proceed to a certain point with limited mass education but then hits a ceiling. Thailand was proceeding up the development ladder in the 1990s, but when the time came to follow the other Asian economies from simple manufacturing into high tech electronics, it simply could not take the next step because only 14 percent of each age cohort had graduated from high school. Understanding this, the Chinese

Figure 7.2
Tertiary School Enrollment

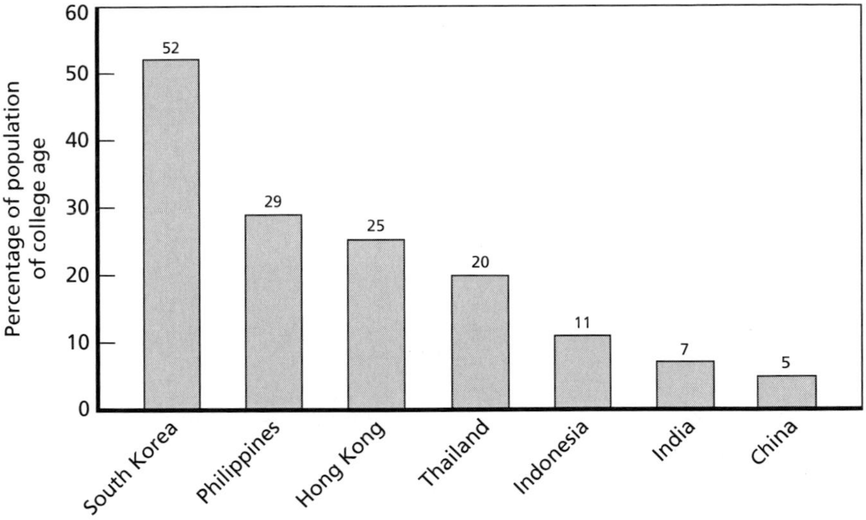

SOURCE: World Bank, World Development Indicators 2002, CD-ROM version.
RAND *MG260-7.2*

government is determined to increase expenditures on education substantially. In fact, the 2003 Budget Report says, "We shall continue to comply with laws and policies which stipulate that the increase in these expenditures [on education and science and technology] should be higher than the growth of the regular revenue."

Public expenditures on health are also relatively low in China, not only compared to developed countries but also to medium-income developing countries. Municipalities have forced clinics to impose fees that cover an appreciable share of costs because the central government has been unwilling to pay for these services. Citizens have turned to private-sector providers because the public sector does not provide adequate care. This will have to change—China's public health care system is in shambles and many private providers are little more than con artists. Families in poorer regions pay a very large share of the costs of education and health care from their own pockets.

In Table 7.1 we project potential future government expenditures on education and health as a share of GDP. We assume that the Chi-

Table 7.1
China's Potential Future Budgetary Obligations (% of GDP)

Category	1978	1985	1995	2000	2005	2010	2015	2020	2025
Pensions	—	—	—	2.6	4.4	4.1	4.4	5.1	5.8
Education and Health	4.1	4.6	3.0	4.9	5.9	6.6	7.2	7.9	8.5
Interest					1.6	1.5	1.6	1.7	1.9
Total			6.9	7.5	10.6	11.4	12.5	14.1	15.5
Percentage-point increase compared with 2000					3.1	3.9	5.0	6.6	8.0

nese government will raise public spending in these categories as a share of GDP toward levels characteristic of wealthier developing countries. Under these assumptions, public spending on education and health rises from 4.9 percent of GDP in 2000 to 8.5 percent in 2025.

Future Expenditures on Pensions
Most of China's pensions have been funded on a cash basis by state-owned enterprises, government agencies, or local governments. This system is no longer viable, especially the portion funded by state-owned enterprises. Now that state-owned enterprises have to compete against imports and private businesses in a market environment, they no longer generate the profits needed to pay pensions to former employees. In a number of instances, loss-making enterprises have been closed. Because of the failure of state-owned enterprises and in some cases municipalities to make the requisite contributions, the central government has had to step in to ensure pensioners are paid. The government has also increased pension payments to its own former employees, some of whom were pressed into early retirement. Payments have risen accordingly (Figure 7.3).

The problem of funding the pensions of retired workers from China's state-owned enterprises is bound to worsen. The number of employees in state-owned enterprises is falling whereas the number of pensioners SOEs are supposed to support is rising. In the future, most people will be employed in the nonstate sector, not by state-owned en-

Figure 7.3
Pension Payments, RMB Billions (nominal)

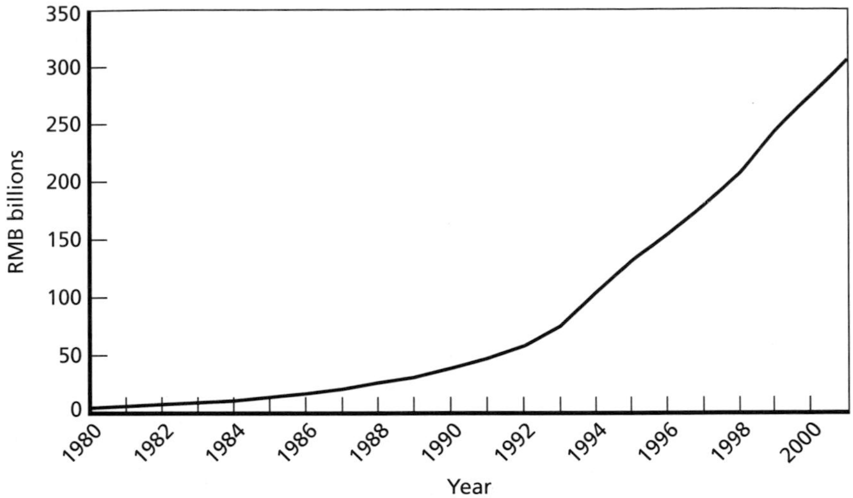

SOURCE: *China Statistical Yearbook 2003*, p. 842.
RAND *MG260-7.3*

terprises. The remaining state-owned enterprises will in most instances not be able to pay the promised pensions.

The Chinese government has already begun to make changes in the pension system to address these problems. During the course of the 1990s, municipalities, provinces, and enterprises began to pool resources. In some areas, standardized contribution rates were introduced across contributing institutions and enterprises. Despite these changes, the current hodgepodge functions poorly. Contribution rates and payments vary geographically and by employer. Coverage is still confined to state employees; rural inhabitants and employees of private companies, collectives, and TVEs have no national plan. Pensions are paid from current contributions. The system still relies heavily on contributions from state-owned enterprises, many of which have stopped making payments.

The national government has advocated and the provincial government of Liaoning has experimented with introducing a new three-tier system comprising (1) a small base pension funded from a payroll tax and guaranteed by the government; (2) mandatory individual accounts in managed investment funds funded by wage taxes paid by the employer and the employee; and (3) tax preferences for voluntary contributions to private retirement accounts. Individuals and employers would contribute a set amount to the mandatory individual accounts, but the benefit would depend on the rate of return earned by the invested funds. These types of schemes have become common in many other transition economies such as Hungary, Kazakhstan, and Poland, where they have worked surprisingly well, although the investment fund managers must be carefully vetted.

China's pension challenges will have major fiscal implications, limiting the availability of funds for other purposes, including military spending. Currently, China has no national state-run program; in particular, people in rural areas are not covered by the national system. Rural inhabitants and other individuals not covered by the state system rely on savings or their children, especially their sons, for financial support when they become too old to work. In the past, villages and local communities stepped in to take care of the childless elderly and the handicapped, but these local support networks have frayed in many areas as individuals have had to become more self-sufficient economically and population movements and urbanization have weakened ties to neighborhoods and villages. As China becomes wealthier (and older), political pressures to provide universal pensions to the elderly will rise, especially as a growing number of elderly couples do not have sons. Consequently, we believe some sort of universal national system is inevitable.

To illustrate the fiscal implications of the Chinese government providing a minimum base pension to the entire population of the elderly, we have constructed the following projection. We assume that the Chinese government provides a universal base pension benefit equivalent to 20 percent of average Chinese wages to all citizens 65 years in age or older (see Table 7.1). Under this assumption, the base pension in 2001 would have been RMB 181 a month, or $22 at the official exchange rate. We assume the average wage and hence pension rises in line with growth in per-capita GDP. We also assume that this portion of the system remains pay-as-you-go. Using the projections of economic growth from Chapter Two, under these assumptions we forecast total pension costs under a universal system rising from 3.1 percent of GDP in 2005 to 5.1 percent in 2025. This figure is only for creating a universal system; it does not include the costs of making up arrears from the current system.

These projections are based on only one policy alternative: providing a small base pension for all China's elderly. However, the Chinese government has a number of policy options available to it. Some of these would entail even higher costs. For example, the Chinese government might choose to provide a more generous level of benefits. It also might encourage the development of self-financing systems through tax incentives or fiscal support.

The government would absorb transition costs, if it should choose to move from the pay-as-you-go system toward a funded system. Following the introduction of a funded system, tax revenues that used to be channeled directly from wage earners to pay current pensioners would be invested in investment accounts that will fund future pensions. Although this shift is a healthy one, costs can be quite substantial. The IMF estimates that in the case of China these transition costs would run 10–15 percent of GDP just to move from the current partial system to a funded system.[1] World Bank studies have generated estimates of the cost of fully capitalizing the current system that run from about half to 100 percent of GDP.[2] However, fully capitalizing the system immediately (covering future pension costs by investing in a dedicated account today) would be highly unusual. This process is usually spread out over a number of years. Moreover, most state-funded pension systems are not fully capitalized; they continue to fund some payments from current revenues.

The cost of capitalizing all or part of the system should be considered part of government debt—amortized over some period of time

---

[1] Interview with Steve Dunaway, China office, IMF, Washington, D.C.

[2] For published estimates by the World Bank of 46–69 percent and of 71 percent, see notes 35 and 36 of Lin, 2003. The figure of 100 percent of GDP was provided orally by a participant in a World Bank survey of several provinces. Lin also cites private studies (notes 43, 44) that came up with fiscal liabilities equal to 28 percent and 41 percent of GDP respectively. Perhaps the most thorough World Bank study of this issue is Yan et al., 2001. That study arrives at an estimate of 71 percent of GDP for the transition cost.

through the budget. The government also faces unfunded pension liabilities from the current system that it will almost certainly cover. Using the IMF's 15 percent of GDP figure, the cost of making up this shortfall, amortized over 20 years (2005–2025) at a real interest rate of 3 percent per year, would be on the order of 1.4 percent of GDP in 2005. This share would decline as the economy grew and the costs were amortized through 2025. Using the World Bank's 50 percent of GDP figure, the costs are significantly higher, running 4.6 percent of GDP in 2005 and then tapering off as the shortfall is amortized. However, because it is not clear whether China is likely to move toward a funded system, we have not included these costs in the projections shown in Table 7.1.

Future Costs of Servicing Government Debt Stemming from Unfunded Liabilities

In Table 3.7 we provided estimates for some of the major liabilities, funded and unfunded, facing the Chinese government. These include official government debt (16 percent of 2003 GDP), recapitalization bonds and government debts to international financial institutions (9 percent), local government debt likely to devolve to the national government (9 percent); and nonperforming loans written down by 80 percent of face value (40–60 percent of GDP), among others. Counterbalancing these claims are liquid government assets that could be used to cover these liabilities (state-owned enterprises that could be privatized and excess foreign reserves) estimated at 25 percent of 2003 GDP, for a net obligation of about 60 percent of GDP. As noted above, however, other potential obligations, including transitioning to a funded pension system if the Chinese government should choose to take this route, and covering pension arrears would add to this total.

Currently, this net debt is financed by imposing implicit costs on Chinese savers, by running down the capital and reserves of state-owned banks, by adding to the liabilities of the central bank, and by directed borrowing to enterprises to cover interest obligations and continued losses. When these obligations are recognized, they will result in a one-time increase in China's national debt. Moreover, as financial markets are liberalized, competition for deposits will push up interest rates, forcing the government to pay market rates on its domestic bonds.

Table 7.1 shows our projection of the interest costs of government debt, incorporating these liabilities. Because China's national debt is likely to be sustainable even after adding these additional obligations, we assume that the debt is not amortized over time but that, after restructuring the financial system, China experiences a permanent increase in the debt stock. We assume that the recognition of these unfunded liabilities is a one-time operation: After the banking system is recapitalized, the system will operate on a commercial basis. The Chinese government will not permit banks to accrue another large stock of nonperforming loans. We also assume that pressures for increased expenditures and limitations on increasing tax revenues will result in manageable fiscal deficits of 3 percent annually over the next two decades, leading to a sustained increase in government debt. We calculate interest costs on this debt under the assumption that after financial market liberalization, the Chinese government will pay a real interest rate of 3 percent on its debts.

Projections of Government Expenditures on Pensions, Education, Health, and Government Debt Service

Table 7.1 also shows our projections of future government spending on a basic pension, publicly provided health care, education, and the interest costs of Chinese government debt. Combined, spending on these items is projected to rise from 7.5 percent of GDP in 2000 to 15.5 percent in 2025, an increase of 8.0 percentage points. This is a very substantial increase in the share of output flowing through government hands.

How realistic are these projections? To some extent the increases in government expenditures of education and health as a share of GDP represent shifts in financing from households to the government. As noted above, Chinese citizens pay a substantial share of both educational and health costs out of their own pockets. These expenditures are inequitably spread across the country. Rural and poorer households often pay a greater share of education and health costs themselves than do richer, urban households. The latter live in provinces that enjoy higher tax revenues where the local governments have the wherewithal to pay for a larger share of these costs. Greater national expenditures on health and education would reduce the current inequities in the delivery of government services, relieving poorer residents of costs for education and health that they currently pay out of pocket. This said, the projected increases are substantial. Under these assumptions, government expenditures as a share of GDP would rise by almost a third, a very large increase.

Other Costs

In the projections above, we did not include all categories of government expenditures that are likely to rise in the coming years because the uncertainties concerning future expenditures in these categories are even greater than those for which we have constructed projections. However, political pressures to increase expenditures in these categories are real and will constrain the Chinese government's ability to increase expenditures in other categories, including military spending.

Unemployment. With tens of millions of workers being laid off and further tens of millions migrating to the cities to look for work, China needs a social safety net for the unemployed and indigent. It is trying to build one. In 1999, most laid-off workers from state-owned enterprises or government offices (93.3 percent) received a basic living allowance averaging RMB 129 per month (about $15, almost below subsistence level).[3] The cost of this support is rising fast. Between 1998 and 2002, the number of urban dwellers receiving subsistence allowances rose from 1.84 million to 20.6 million people.[4] As the "iron rice bowl" rusts, the cost of providing support to the unemployed will continue to rise.

We have made no explicit attempt to project future costs for unemployment insurance and welfare here. The variety and cost of potential support programs governments are willing (or unwilling) to provide are so great that we have not included specific forecasts of these costs. However, we note that political pressures for government support of

---

[3] Hu, 2002, p. 16.

[4] Xiang, 2003.

those losing their jobs will almost inevitably rise as China becomes richer and the costs of adjustment continue to shift from rural areas to the politically more important urban areas.

The Environment. China is not just at risk of an environmental crisis, it is having an environmental crisis. Virtually every year brings serious floods and droughts caused by deforestation and other environmental problems. The air in most of China's major cities is a serious health hazard; the World Bank maintains that excessive air pollution alone is costing 289,000 lives per year as well as 7.4 million person-years of work lost to illness.[5] A vast swath of northern China threatens to become another dustbowl. Beijing's water table drops measurably each year.

The extent of environmental damage is so great that a number of Chinese citizens and government officials are becoming environmentally conscious. The Chinese government has developed a 50-year plan for the environment. However, the construction of a system for administering and enforcing environmental regulations is in its infancy. The World Bank estimates that China needs to spend 2.1 percent of GDP per year to clean up its air and 1.1 percent of GDP to clean up its water.[6] Based on past performance, not all these expenditures would come out of the government's budget; many would be imposed on companies or be deferred.

Income Equalization: Transfers of Resources from Richer to Poorer Regions

As noted in Chapter Three, one of the prices for the tax reform of 1994 was a sharp reduction in transfers from rich provinces to poor ones to give all provinces an incentive to collect additional taxes, because most of the incremental revenues would accrue to the provinces themselves. To ameliorate inequalities, the government will have to reduce the share of revenues that the richer provinces are allowed to retain and transfer those revenues to poorer provinces by providing disproportionate support for national programs such as education. These transfer issues differ in principle from other budget expenditures because they shift

---

[5] See World Bank, 1997, p. 73.

[6] See World Bank, 1997, Table 6.2, p. 80.

resources from one set of individuals or institutions to another; they do not involve government-dictated purchases of goods and services. But the sums are large, and the political effort required to fund such programs is no different from the political and administrative efforts needed to raise funds for other programs. Moreover, such a policy shift will reduce the incentives of richer provinces to increase official tax collections. Thus, these political pressures will also affect Chinese government decisions on expenditure decisions, including expenditures on the military.

Future Growth in Total Government Expenditures

In Chapter Two, we argued that economic growth in China will slow over the course of the next two decades as many factors that drove economic growth since 1978 diminish. Productivity growth in agriculture, which continues to be a major employer, has virtually ceased because the easy gains from improving incentives and rationalizing prices have already been taken. The sheer size of China's international trade will limit the extent of further gains from trade. China has already experienced three—1998, 1999, and 2001—when growth in trade slowed sharply, although growth rates have since rebounded. Continued problems in the financial sector, especially in the allocation of investment, will constrain growth in capital productivity. The aging of China's population and incipient declines in the size of the labor force will curb growth from increases in the supply of labor. Consequently, RAND forecasts that economic growth will slow to 7 percent per year through 2010, gradually declining to 3 percent per year in 2025, for an average annual rate of 5 percent through 2025, substantially lower than the average annual rate of growth of 8.7 percent reported for the past quarter-century. However, even at these rates of growth, the Chinese economy will become much larger, more than tripling in size by 2025.

As noted in the previous section, the political pressures for increased public services are so strong and the current provision of education, health care, and pensions so inequitable and dysfunctional that it will be extremely difficult for the Chinese government to restrain growth in public expenditures on these and other government services in the years ahead. Consequently, government expenditures appear destined to grow more rapidly than the projected rate of growth in GDP. To increase overall government expenditures at a rate faster than that of GDP, the Chinese government has three options: (1) It can cover increased expenditures by permitting the budget deficit to widen, borrowing more from home or abroad to cover the gap; (2) it can print more money to cover the deficit, thereby financing the deficit through an "inflation tax," (3) it can increase the share of GDP taken in taxes or fees.

As Table 7.2 shows, current government budget deficits are approximately 3 percent of GDP. The Chinese government finances these deficits by borrowing domestically. If the Chinese government were to permit the deficit to widen, it would have to borrow more heavily at home or abroad. Although China is a country of high savers, increasing the budget deficit by a percentage point of GDP or more raises a number of dangers. First, as noted above, the Chinese government has latent liabilities that when recognized will boost government debt substantially. Increasing the size of the operating deficit during a period when government debt is rising sharply could well contribute to a fall in confidence, resulting in an increase in interest rates or a fall in the value of the renminbi as domestic and foreign creditors take fright and move their savings abroad. Developing countries in Asia and elsewhere have suffered from balance-of-payments crises brought on by a collapse

Table 7.2
China's and Developing Country Budgets as a Percentage of GDP

	China's Official Budget (2002)	China's Official Budget plus Off-Budget (2002 est.)	Middle-Income Countries (1998)	Low-Income Countries (2001)	World
Expenditures	21.7	25.7	23.1	18.0	25.7
Revenues	18.7	23.0	18.8	13.9	24.7
Balance	−3.0	−2.7	−4.3	−4.1	−1.0

SOURCE: China Statistical Yearbook; World Bank, World Data Profile, http://www.worldbank.org/data/countrydata/countrydata.html.

in confidence. If government deficits are permitted to rise, China is likely to suffer a similar fate.

The second option, financing increased government spending by printing money, has already been rejected by China's leadership. Monetary finance of deficits invariably generates inflation. The Chinese leadership has found past bouts of inflation politically dangerous. In fact, the hyperinflation of the Kuomintang period is considered a major reason for the demise of that government. After the last several years of low inflation, a surge could well be accompanied by organized protests, strikes, and possibly violent demonstrations that would challenge the Chinese Communist Party's hold on power. In the recent past when food prices have surged, violent demonstrations have flared.

The third option is to increase the share of government revenues in GDP by raising taxes or increasing fees. During the 1980s, total government revenues, on and off budget, averaged 37 percent of GDP. The government's take declined precipitously between 1983 and 1995 as the government introduced a new tax system and moved responsibility for investment from ministries to enterprises. Revenues have rebounded since 1995 as the government extended the value-added tax and corporate taxes (Figure 7.4). By 2002, revenues as a share of GDP had risen to 18.7 percent, up 8 percentage points from 1995, reflecting a return to a more normal state of affairs. If one adds in estimated off budget revenues, this ratio rises to an estimated 23 percent of GDP.

Although this increase in government revenues as a share of GDP over the past eight years has been dramatic, further increases will be harder to come by. Tax revolts are already common in rural areas. Taxes and government expenditures as a share of GDP are now very close to the average shares in other developing countries (Table 7.2). Although some developing-country governments extract and spend a higher percentage of GDP than China does, these countries often pay a price in slower growth as resources are diverted to support large government bureaucracies or squandered on inefficient benefits schemes.

The Chinese government is likely to have some difficulty in substantially boosting the share of output extracted through taxes or government-imposed fees in the coming two decades. It was able to extract large shares of economic output in the 1980s because most economic

Figure 7.4
China's Tax Revenues and Government Expenditures as a Percentage of GDP

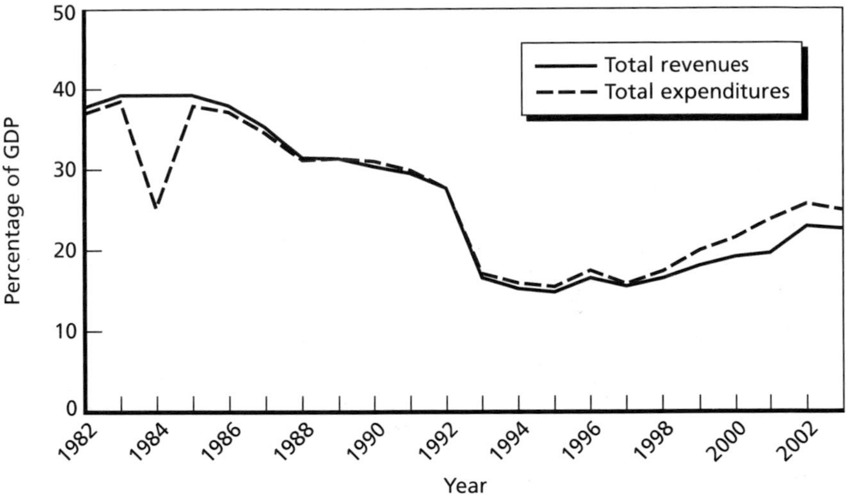

RAND *MG260-7.4*

activity was still generated by the state-owned sector. Although managers of state-owned companies attempted to hold onto revenues, the Chinese Communist Party and the government were powerful enough to compel managers to pay taxes on profits and revenues. Moreover, the high (37 percent) share of GDP extracted by the government during these years was somewhat illusory. During this period the Chinese government was responsible for providing funds for investments by state-owned enterprises, so that many of the funds extracted from the state-owned sector returned to the selfsame enterprises for investments. Because these revenues were needed for investment, the government's ability to divert these funds for expenditures on public services was limited.

Increasing the share of output taken through taxes or fees will necessitate a major shift in the tax burden. Now that more than half of economic output is generated by the nonstate sector and many state-owned enterprises lose money, the government will have to rely much more heavily on taxes from the nonstate sector. Taxing the nonstate sector is substantially more difficult than extracting revenues from the state-owned sector. As noted in Chapter Three, tax evasion is pervasive in China. Many private businesses consist of sole proprietorships, which are inherently difficult to tax because owners are able to conceal revenues by conducting transactions in cash or to reduce profits by inflating costs. Even large businesses have become adept at tax avoidance. According to Chinese press reports, bribery of tax officials, false accounting, and hiring employees off the books are endemic. Eliminating these practices implies wholesale reform of the government administration.

The government will also need to shift taxes from businesses to households. Currently, businesses bear the brunt of the tax burden in China. Taxes on factors of production (labor and capital) are relatively small. The shift from taxes on businesses to households can take place either in the form of new or increased consumption taxes, income taxes, or property taxes. However, in China's economic environment, none of these taxes is easy to levy. Despite the existence of an obligation to pay individual income taxes, personal income tax evasion is widespread. The government has been more successful at collecting consumption taxes like value-added tax (VAT), which accounts for about one-third of all tax revenues. However, small private businesses evade this tax as well.

China will also need to shift the tax burden from rural to urban households. As noted in Chapter Two, the gap between urban and rural incomes is the greatest among larger developing countries and is widening. Despite this gap, Chinese citizens living in rural areas tend to pay a higher share of their incomes in taxes and fees than do citizens of urban areas, including substantial fees for education and health care, services which are more heavily financed by local governments in urban areas. The fees and taxes that rural households are paying already elicit frequent protests and even outbursts of violence. Recognizing these inequities, the Chinese government has been working to reduce the tax burden on rural inhabitants. In 2002, a reform of taxes and administrative charges in rural areas was extended to 20 provinces, reportedly reducing the aggregate tax burden on farmers by an average of 30 percent.[7] However, reducing the tax burden on rural households requires increasing taxes on urban households. As labor continues to

---

[7] Xiang 2003, p. 3.

shift from state-owned enterprises to the private sector in urban areas, increasing the tax burden on urban residents will not be easy.

In short, over the course of the coming two decades, demands for increased government expenditures on education, health, and pensions, let alone pressures for increased expenditures on the environment, agriculture, and public infrastructure, especially in rapidly expanding urban areas, will push government expenditures up. Because of these pressures for spending and continued difficulties in raising taxes, the Chinese government will face a fairly stringent budgetary outlook despite continued strong economic growth. The pressures are so severe and cover so many areas vital to political stability and continued economic growth that, notwithstanding the uncertainties involved in forecasting future spending demands, this conclusion is extremely robust.

Future Chinese Military Expenditures

Since the PLA put down the demonstrations in Tiananmen Square in 1989, a number of indicators suggest that the Chinese government has significantly raised the priority given military spending. Although, as shown in Chapter Four, the official defense budget does not include all military spending, the official budget does appear to capture expenditures on a number of major items and therefore is a useful metric for flagging changes in government priorities. After sharp reductions in reported military spending, both in real terms and as a share of GDP in the 1980s, officially reported military spending grew at double-digit rates between 1989 and 2002, reflecting the enhanced importance of the military. Although this string of increases was broken in 2003 when growth returned to single digits as the official military budget rose 9.4 percent, the official budget still rose more rapidly than growth in overall government expenditures, which was kept to 7.7 percent in that year.

Other indicators also show that the Chinese government has given expenditures on the military a higher priority in the past decade. Although the number of people in uniform has fallen, military pay and benefits have risen sharply. Since the mid-1990s, China has increased expenditures on military procurement to purchase a number of more capable combat aircraft and naval ships, many of which have come from abroad.[8]

As a consequence of the higher priority given military spending by the Chinese government, the share of GDP taken by the official defense budget has increased from a low of 1.06 percent of GDP in 1996 to 1.66 percent in 2003. According to our own estimates of spending derived in Chapter Four, total expenditures fell into a range from 2.3 to 2.8 percent of GDP in 2003. All three of these estimates (1.66, 2.3, and 2.8 percent of GDP) are not extraordinarily large as a share of economic output. During the 1980s, most NATO and Warsaw Pact countries (other than the United States and the Soviet Union) spent from 2 to 3 percent of GDP on the military. Consequently, the Chinese government has some room to increase military expenditures as a share of GDP.

That said, recent rates of increase in military spending are not sustainable. Between 1988 and 2003, officially reported spending on defense rose 9.8 percent per year on average in real terms. If the Chinese government should attempt to keep defense budgets growing at 9.8 percent per year while the economy grows at our projected rates, military expenditures would take between 6.2 and 7.6 percent of GDP by 2025, starting from our lower and higher estimates of 2003 spending, respectively. In other words, the share of military spending in GDP would rise 2.5 times over this period.

In our view, it would be extraordinarily difficult for the Chinese government to increase the share of GDP devoted to the military to these levels. Since the end of the Cold War, no government has sustained military expenditures of 6 percent or more of GDP, unless it was at war or it perceived an eminent threat to its national security as Israel does. As noted above, China faces very strong pressures to increase government expenditures in nonmilitary categories. We estimate that raising public expenditures on education and health, introducing a very modest universal pension system, and converting unrecognized government liabilities into government bonds and paying interest on this debt are likely to result in an increase in government expenditures as a share of GDP of 8 percentage points, compared to the year 2000.

---

[8] IISS, 1997, 2003.

In light of pressures to raise spending in these categories, increasing expenditures on the military by an additional 4 to 5 percentage points of GDP would be very difficult. In the face of all the other budgetary pressures facing the Chinese government, the Chinese political and military leaderships would have to perceive a much more immediate threat to China's national security than indicated by our analysis in Chapter Six to absorb the political costs of raising military expenditures to over 6 percent of GDP in the next two decades. In short, we argue China will not be able to afford to continue to increase military spending at the pace of the last few years.

Projections of Future Military Expenditures in 2001 Renminbi. Assuming that China's current security environment does not worsen, what levels of military spending are likely to be supportable over the next two decades? To answer this question, we generated two forecasts of potential future military expenditure levels. Both forecasts start with the assumption that the projections shown in Table 7.1 are indicative of future expenditure patterns: The Chinese government will choose to substantially increase expenditures on education, health, and pensions and will be compelled to convert currently unrecognized liabilities into government debt and pay market interest rates on its debts. The forecasts also assume that the Chinese government will be able to extract a higher share of output through taxes and fees to pay for this spending, primarily through more effective tax policies and tax administration. Increased taxation is assumed to be politically possible because some of the increase in government expenditures will involve replacing current private spending on education, health care, and retirement by government spending.

In addition to these forecasts for increased spending, we assume that the Chinese government is able to gradually eliminate price subsidies and government financial support to state-owned enterprises. Spending on administrative costs is assumed to remain relatively constant at 3.7 percent of GDP. However, despite China's needs for infrastructure, government investment in infrastructure is assumed to fall from 6.75 percent of GDP to 5.2 percent by 2025 as it is crowded out by demands for more social spending and the government increasingly turns to privately funded projects to ease its budgetary problems.

In our high-end scenario, military spending is projected to rise from our upper-bound estimate of 2.8 percent of GDP in 2003 to 5.0 percent in 2025. In our view, the 5.0 percent of GDP figure is a maximum. We have already described the enormous nonmilitary spending pressures facing the Chinese leadership in the next two decades and the difficulties it faces in increasing tax revenues as a share of GDP. The experiences of other developing countries in recent decades suggests that 5 percent of GDP is very much at the high end of what governments are able to devote to their militaries, unless they perceive a clear and present danger. Figure 7.5 shows military spending as a share of GDP for a number of Asian developing countries. As can be seen, with the exception of Pakistan, none of the Asian countries spends even 3 percent of GDP on its military. India, a country that has almost gone to war in recent years, spends 2.7 percent of GDP. Pakistan, which is run by a military dictatorship and is involved in conflicts on its borders with India and Afghanistan, is the only outlier. It spends 4.4 percent

Figure 7.5
Asian Military Budgets as a Share of GDP in 2002

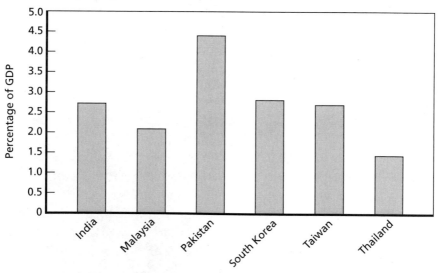

SOURCES: Defense Budgets: Institute of International Security Studies, The *Military Balance*, 2003-2004; GDP in domestic currency: IMF, *International Financial Statistics*, April 2004.

RAND *MG260-7.5*

of GDP. Expenditures in Latin America are much lower than in Asia. Outside the United States, which spent 3.3 percent of its GDP on the military in 2003, no NATO country spends more than 3.0 percent of GDP on the military; most spend less than 2 percent. During the late 1990s, even U.S. military spending fell below 3.0 percent of GDP.

Under these assumptions, total budgetary expenditures rise to over 33 percent of GDP, well above the average expenditure share of GDP of 23.1 percent characteristic of middle-income developing countries. Government revenues from taxes and fees would have to rise to 30 percent of GDP to finance this level of expenditures, assuming that government budget deficits have to be kept to 3 percent of GDP or less to ensure that macroeconomic stability is maintained. In our view, this projection is very much an upper bound concerning aggregate Chinese government spending and revenues. It incorporates the probable maximum share of the economy that the Chinese government would be able to extract in taxes and fees.

For our second projection we take our lower bound estimate of total Chinese military spending in 2003 of RMB 257 billion (2.3 percent of GDP) and assume that other budgetary expenditure pressures and constraints on the ability of the government to increase taxes limit overall expenditures to 30.5 percent of GDP in 2025. In this case, we assume military spending stays constant at 2.3 percent of GDP through 2025.

Because of China's continuing solid rates of economic growth, military spending rises very sharply in real terms by 2025 under both projections (Table 7.3). In the case of our upper-bound projection, military spending rises to be 5.2 times as large in 2025 as in 2003. In the mid-range projection, military spending still rises, but more modestly: in real terms it almost triples by 2025. By 2025, the higher projection is 2.2 times the lower.

Because of the many competing demands on government spending and the difficulties the Chinese government is likely to face increasing taxes, we believe it is highly unlikely that the Chinese government would expend 5 percent of GDP on the military unless it perceives a much higher threat to its national security than is currently the case. Consequently, we argue the mid-range forecast is the more plausible.

Table 7.3
RAND Projections of Chinese Military Spending Through 2025

	Billions of 2001 Renminbi				
	2003	2010	2015	2020	2025
Maximum projection	314.5	702.7	1001.7	1307.4	1633.3
Mid-range projection	256.6	416.6	531.7	646.9	750.0
	Percentage of GDP				
Maximum projection	2.8	4.0	4.3	4.6	5.0
Mid-range projection	2.3	2.3	2.3	2.3	2.3
Ratio between maximum and mid-range	1.23	1.73	1.88	2.02	2.18

Dollar Projections of Future Military Expenditures at Market Exchange Rates. How do these expenditures compare with those of other major powers? Table 7.4 converts our estimates of year 2003 and projections of future Chinese military spending in 2025 in renminbi into 2001 dollars using the official average renminbi/dollar exchange rate in 2001. We then compare these figures with military expenditures in dollars by France, Japan, Russia, the United States, and the United Kingdom, who, with China, comprise the top six countries in the world in terms of military expenditures. As can be seen, our low-end estimate of $31 billion puts China at the bottom of this group. Our high-end estimate for China is still the lowest of the group, although it is similar in size to the United Kingdom's expenditures. Only the United States and Russia spend substantially more than China on their militaries. The 2003 U.S. defense budget is 14 times the official Chinese budget and over eight times our upper-bound estimate. However, the Russian estimate is not strictly comparable as it is based on Western estimates, which utilize purchasing power parity exchange rates to convert some expenditures into dollars.

Using the market exchange rate of 2001 to convert projections of Chinese military spending in 2001 renminbi into 2001 dollars, our high-end projection of future military expenditures rises from $38 billion in 2003 to a projected $197 billion in 2025 (Table 7.4). In our mid-range estimate, military expenditures increase from $31 billion to $91 billion over the same period. For illustrative purposes, we have gen-

Table 7.4
Current and Future Chinese Military Expenditures and Current Military
Expenditures by Other Major Powers

	Billions of 2001 Dollars at Market Exchange Rates		
Country	2003	2025	Assumed Average GDP Growth Rate
RAND maximum for China	38.0	197.3	5.0
RAND mid-range estimate for China	31.0	90.6	5.0
China official	22.4	65.4	5.0
France[a]	40.2	69.4	2.4
Japan[a]	39.5	62.3	2.0
Russia[a]	50.8	125.2	4.0
United States	304.7	583.9	3.0
United Kingdom[a]	37.3	67.5	2.6

[a]2002 data.
SOURCES: France, Japan, Russia, and the United Kingdom: IISS, The Military Balance, 2003–2004; United States Department of Defense, Office of the Undersecretary of Defense, Comptroller, National Defense Budget Estimates for 2004, March 2003, http:// www.dod.mil/comptroller/defbudget/fy2004/fy2004_greenbook.pdf; China: RAND estimates. Projected GDP growth rates: China and Russia: RAND estimates; all other countries: International Energy Outlook 2002, Washington, D.C.: U.S. Department of Energy, Energy Information Agency, March 2002, p. 182.

erated very rough projections of future military spending by the other powers assuming that military expenditures in these other countries will rise at the same rate as their projected GDP. The GDP forecasts were taken from U.S. government publications or other RAND work. Under the high-end projection, China's military expenditures will substantially exceed those of any other power except the United States by 2025. U.S. expenditures would remain almost three times larger, however. Under the mid-range projection, expenditures would exceed those of the West European powers and Japan by about one half. However, Russian expenditures would continue to exceed those of China even in 2025, and U.S. expenditures would still be six times larger.

Dollar Projections of Future Military Expenditures Using a Combination of Market and PPP Exchange Rates. In Chapter Two, we described the problems of converting renminbi estimates of GDP into dollars using market exchange rates or purchasing power parity exchange rates. In Chapter Four we discussed some of the perils of using market or PPP rates to translate military expenditures in renminbi into their dollar equivalents. On the one hand, market exchange rates tend to underestimate the purchasing power of renminbi expenditures for goods and services purchased by the Chinese military that are not traded internationally. For example, military personnel costs are much lower in China than in the United States; consequently, renminbi expenditures go much further in terms of covering wages and living costs of military personnel than dollar expenditures calculated at market exchange rates would suggest. In these expenditure categories, PPP exchange rates provide a better reflection of the value of the nontraded goods and services purchased than do market exchange rates.

On the other hand, PPP exchange rates tend to exaggerate the purchasing power of renminbi expenditures for such tradable high technology goods as military equipment. As noted in Chapter Five, Chinese manufacturers of more sophisticated military equipment appear to face higher rather than lower costs of production than U.S. manufacturers. These enterprises have to compete with the Chinese private sector for the highly educated labor needed to develop and produce quality weapon systems and must pay market prices to procure the sophisticated components, machinery, and equipment needed to manufacture these systems. These inputs are imported or sold at world market prices by domestic Chinese producers. Thus, market exchange rates are more appropriate for converting renminbi expenditures on military procurement.

In Chapter Two, we attempted to provide a better estimate of China's GDP in dollars by converting renminbi output into 2001 dollars employing market or PPP exchange rates for those sectors of the Chinese economy for which they are best suited. This method captures the best features of both methods. We first roughly divided China's economy into sectors that produce tradable goods and services and those that do not. We then converted output from sectors producing tradable goods into 2001 dollars using the market exchange rate. We converted output of nontraded goods and services into 2001 dollars using the PPP exchange rate. To forecast China's future GDP in dollar terms, we first projected output by sector (tradable and nontraded goods and services). We then forecast likely changes in the real effective market and PPP exchange rates over time. These exchange rate forecasts are driven by projected structural changes in China's economy. We then used these projections of sectoral output growth and changes in the market and PPP exchange rates to convert our forecasts of China's GDP in renminbi into constant 2001 dollars. We believe these estimates avoid the overexaggeration inherent in forecasting dollar GDP using today's PPP exchange rate and the underestimation inherent in using today's market exchange rate.

Below, we use a similar technique to provide estimates and projections of Chinese military expenditures in 2001 dollars that better reflect the purchasing power of China's military budgets compared with U.S. defense budgets. We first disaggregate and project expenditures by category in renminbi: personnel, operations and maintenance, and procurement. We begin with breakdowns by these categories from the official Chinese military budget provided in the 2002 *Defense White Paper*. We then break down the RAND estimates of additional expenditures hidden in the rest of the budget, generated in Chapter Five, into these three categories and add them to the official numbers. For example, funds for foreign arms imports, defense industrial subsidies, defense R&D, and arms sales revenues are grouped with procurement whereas local support for defense and expenditures on paramilitaries are grouped with expenditures on personnel. Revised expenditure shares for personnel, operations and maintenance, and procurement and research and development are then used to create an estimate of the actual composition of military spending in 2003. Based on the analysis of desired future military capabilities in Chapter Six, we then posited a future composition of military spending in the domestic currency, the renminbi, which is closely aligned with those of modern militaries: expenditure shares of roughly one-third on each category. Spending is assumed to gradually move toward this desired end-state. We then projected future renminbi military expenditures by category by multiplying these shares times the projected budget for each year.

Subsequently, spending in each category is converted into dollars using the most appropriate exchange rate. Expenditures on personnel are converted into 2001 dollars using our projections of future PPP exchange rates because those costs primarily consist of purchases of nontraded goods such as housing, personal services, or wages. Expenditures on operations and maintenance and procurement and research and development are converted into 2001 dollars using our projections of future market exchange rates, as military equipment is either imported from abroad or manufactured from materials and components whose prices are dictated by world markets. This approach overcomes some of the valuation problems inherent in attempting to compare military expenditures in a common currency.

Using this methodology, for 2003 our high-end estimate runs $75.6 billion in 2001 dollars and our mid-range estimate, $68.6 billion in 2001 dollars. The high-end figure is about 17 percent above the high end of the most recent Department of Defense estimates of PLA funding, $45 billion–$65 billion; our mid-range estimate is just above this range.[9] Personnel costs for both the higher and lower estimates, when converted to dollars at the purchasing power parity exchange rate, ran $49 billion in 2003 (Table 7.5). Because personnel costs are valued using PPP exchange rates and operations and maintenance and procurement costs are valued using the lower market exchange rates, 65 percent of our high-end estimate of Chinese military expenditures for 2003 and 71 percent of the mid-range estimate consist of personnel costs. Our high end estimates of expenditures on procurement runs $10.5 billion in 2003, which would be almost double similar expenditures by other major powers, excluding Russia. According to these estimates, Chinese expenditures on procurement are very substantially more than Russia's, which has slashed procurement spending in recent years.

Table 7.5 also shows our projections of future Chinese military expenditures converted into dollars using this combination of projected market and PPP exchange rates. Our high-end forecast rises from an estimated $75.6 billion in 2003 to $403 billion in 2025. The lower estimate has expenditures rising from $68.6 billion in 2003 to $185 billion in 2025. The size of both estimates for 2003 compared to military spending converted at market exchange rates is driven by the very large disparity between the purchasing power parity exchange rate and the market exchange rate (1 to 4.4).

---

[9] U.S. Department of Defense, 2003.

Table 7.5
RAND Projections of Chinese Military Spending Through 2025: Combined
Market and PPP Exchange Rates (billions of 2001 dollars)

	2003	2010	2015	2020	2025
Mid-range projection	68.6	91.2	113.7	143.9	185.2
Personnel	48.9	57.8	65.0	73.1	82.2
Operations and maintenance	8.6	15.3	23.0	34.6	51.9
Procurement and R&D	11.1	18.1	25.6	36.2	51.1
Maximum projection	75.6	145.0	207.4	287.3	403.4
Personnel	48.9	84.7	111.5	141.0	178.9
Operations and maintenance	8.3	22.3	39.6	67.1	113.0
Procurement and R&D	18.5	38.0	56.2	79.3	111.4
Ratio between maximum and mid-range projections	1.10	1.59	1.82	2.00	2.18

Which of these estimates and projections in dollars—those at the market exchange rate or those projected using a combination of market and PPP exchange rates—best reflect the value of China's military expenditures? From the perspective of the U.S. military, the dollar projections based on a weighted combination of PPP and market exchange rates best capture the value of Chinese military spending compared with U.S. spending. These projections adjust for the greater purchasing power of the renminbi in terms of personnel and its lower value in terms of purchases of sophisticated weapons systems. The procurement and operations and maintenance expenditure estimates and forecasts should, by and large, be equivalent to U.S. expenditures of the same size on these items. These dollar forecasts also explicitly take into account likely shifts in the relative purchasing power of the renminbi in the future: As wages and living costs rise in real terms, the purchasing power of renminbi expenditures on personnel will decline. On the other hand, as the market exchange rate appreciates in coming years, the purchasing power of the renminbi in terms of more sophisticated weapons will rise.

What Might These Future Military Expenditures Buy?

In this section, we translate these projections of future Chinese military expenditures in 2001 dollars into more concrete measures of the resources that China may devote to its military in the future by comparing them with equivalent U.S. expenditures. Table 7.6 compares the projections of future Chinese expenditures for 2025 on personnel, operations and maintenance, and procurement developed above with U.S. expenditures on these items in 2003 in constant 2001 dollars. As can be seen, under the maximum projection, Chinese military spending becomes appreciable by 2025, exceeding the U.S. defense budget for 2003 by close to a third. However, the allocation of spending in this projection remains skewed toward personnel costs because China is assumed to retain a large standing army. Because personnel costs are converted to dollars at purchasing power parity exchange rates and rise appreciably over time as a result of increasing real wages, personnel costs account for 60 percent of total costs as opposed to just 27 percent of the 2003 U.S. budget. In the maximum expenditure projection, the other two expenditure categories measured in constant 2001 dollars equal U.S. expenditures in 2003 on these categories. In the mid-range projections, they run a little less than half of 2003 U.S. expenditures on these categories.

We also assess the potential resources that China may devote to purchasing military assets in the coming two decades. To provide a better sense of the cumulative impact on force structure of the projected defense expenditures on procurement, we compare the projected cumulative totals spent on procurement and research and development

Table 7.6
RAND Projections of Chinese Military Spending in 2025 Compared to DoD Expenditures in 2003

	Billions of 2001 dollars		
	RAND Projections for 2025	DoD, 2003	Ratio
Mid-range projection	185.2	304.7	60.8
Personnel	82.2	81.7	100.6
Operations and maintenance	51.9	113.4	45.8
Procurement and RDT&E	51.1	111.1	46.0
Maximum projection	403.4	304.7	132.4
Personnel	178.9	81.7	219.0
Operations and maintenance	113.0	113.4	99.7
Procurement and RDT&E	111.4	111.1	100.3

Table 7.7
Potential Future Chinese Military Expenditures on Procurement Compared
with U.S. Expenditures

Category	Expenditures (billions of 2001 dollars)	China as a Percent of U.S.[a]
Cumulative U.S. expenditures on RDT&E and procurement 1981–2003	2,712.4	
Maximum cumulative projections of Chinese expenditures on RDT&E and procurement 2003–2025	1,279.7	47.2
Mid-range cumulative projections of Chinese expenditures on R&D and procurement 2003–2025	597.8	22.0
Cumulative USAF expenditures on R&D and procurement 1981—2003	1,039.4	
Maximum cumulative projections of Chinese expenditures on PLAAF procurement 2003–2025	490.4	47.2
Mid-range cumulative projections of Chinese expenditures on PLAAF procurement 2003–2025	229.1	22.0

[a]Because the share of total procurement by the PLAAF is assumed to be the same as that of the USAF, the ratios between total Chinese procurement and PLAAF procurement and the U.S. budget and USAF procurement are the same.

through 2025 with U.S. expenditures on these items in 2001 dollars over the past 22 years. Military capabilities, especially stocks of weaponry, are the result of cumulative spending over time, not just current spending. This exercise provides a measure of what China may be able to spend over this period.

As can be seen from Table 7.7, procurement is appreciable in the high-end case. The cumulative total would be close to half of what the United States spent on military procurement and RDT&E between 1981 and 2003. Under this scenario, no other country besides the United States would rival China in terms of stocks of weaponry. In the low-end projection, procurement spending is still appreciable, but in this case, even after 22 years, China's cumulative expenditures on procurement would be only 22.0 percent of what the United States spent between 1981 and 2003.

Finally, to give the U.S. Air Force a more tangible measure of potential future Chinese expenditures on air assets, we present notional projections of future Chinese spending on research and development and procurement of air assets in Table 7.7 and compare them with past

USAF expenditures in the same categories. As noted in Chapter Six, a key objective of the Chinese military leadership is to develop a more modern and versatile air force that possesses advanced longer-range strike and ground-attack aircraft, improved early warning and air defenses, extended and close air support, and longer-range transport, lift, and midair refueling capabilities. As part of these improvements, the Chinese wish to acquire a fourth-generation fighter. These items are very expensive. To purchase them, the Chinese will need to devote considerable resources to modernizing the PLAAF. Our last projection investigates whether the PLAAF is likely to obtain the resources it will need to achieve these goals.

This projection was generated in the following manner. First, we split projected research and development and procurement expenditures into expenditures by service. To make this comparison as clear as possible, we assume that China spends the same share of its R&D and procurement budget on the PLAAF as the United States does on the USAF. We then cumulate these projections and compare them with the past 22 years of U.S. RDT&E and procurement expenditures on the USAF measured in constant 2001 dollars. In support of these assumptions, we note that the PLAAF's share of the total Chinese defense budget, 31.4 percent between 1950 and 1980 (the only period for which we have data), does not differ markedly from the share of the USAF in the budget of the Department of Defense (28.8 percent).[10] Thus, the assumption that the shares of procurement taken by the two services may be similar does not seem outlandish.

Because the share of procurement by the PLAAF is assumed to be the same as the historical share of USAF procurement in total procurement, the ratios between Chinese expenditures on PLAAF procurement between 2006 and 2025 and USAF procurement between 1984 and 2003 and the ratios for total Chinese procurement and total U.S. procurement are the same. The magnitudes are revealing. In our view, the maximum likely expenditures that China would make on RDT&E and procuring weapons and equipment for the PLAAF between 2003

---

[10] See Chapter Five.

and 2025 would be on the order of $490 billion. This is a large sum of money. However, the cost of the current USAF inventory of weapons and equipment in 2001 dollars and the associated RDT&E to develop these systems is more than twice as much. Using our mid-range projection of military expenditures, PLAAF expenditures on procurement and RDT&E would run $229 billion, 22.0 percent of cumulative USAF expenditures over the past 22 years on these same categories.

CHAPTER EIGHT

Conclusions and Indicators

Will China Have the Resources to Become a Serious Military Threat to the United States?

The purpose of this report has been to assess the likely extent of future resources that will become available to the Chinese military through 2025. To make this assessment, we attempted to answer the following questions:

1. What will be the likely shape and size of the Chinese economy over the next two decades?
2. What types of constraints will the Chinese government face in terms of drawing on increased economic output for spending on the military?
3. What problems will the military face and what possibilities will it have in terms of purchasing the goods and services it desires from the Chinese defense industry?
4. Faced with these constraints and opportunities, what resources will the Chinese armed forces likely have at their disposal over the next two decades?

We found the following:

Growth in Resources
Measured at purchasing power parity exchange rates, China's economy is already the second largest in the world, behind only that of the United States. At market exchange rates, its economy is the sixth largest, behind the United States, Japan, Germany, the United Kingdom, and France.

The Chinese economy is destined to become even larger. We project that China's economy will grow at an average annual rate of 5 percent through 2025, more than tripling in size. The projected rates of growth, 7 percent per year through 2010, gradually declining to 3 percent per year in 2025, are substantially lower than the average annual rate of growth of 8.7 percent reported for the past quarter-century. However, it is more plausible than forecasts employing rates of growth of the recent past. First, officially reported rates of growth appear to suffer from an upward bias. The OECD, World Bank, and Chinese economic research institutes all argue persuasively that official growth rates have been exaggerated. Second, and more important, China faces many constraints that will slow growth over the course of the next two decades: stagnation and eventual decline in the labor force, a fall in domestic savings as the population ages, a slowdown in growth in exports and industrial output as a result of market saturation, weaknesses in the financial sector, and problems in agriculture and rural areas.

To compare China's economic output—and hence the potential resources available for military expenditures—with that of other countries, economic output must be converted from the domestic currency into a common denominator such as dollars. Economists use two types of exchange rates to make these conversions: the market exchange rate and the purchasing power parity exchange rate, an exchange rate estimated by comparing the purchasing power of local currencies for the same basket of goods across countries. Neither of these exchange rates is altogether satisfactory for measuring China's GDP in dollars over time. On the one hand, the Chinese government appears to keep the market exchange rate undervalued by fixing or pegging the renminbi against the dollar, thereby imparting a downward bias to GDP converted at the market exchange rate. If, as forecast, China's economy continues to grow more rapidly than the economies of the more-developed countries, its market exchange rate will *rise* in real effective terms, increasing the value of China's GDP in dollar terms above and beyond growth alone.

On the other hand, the purchasing power parity exchange rate is heavily influenced by the low cost of services, housing, and basic foodstuffs in China. As incomes rise, so will the cost of these goods and services produced with local labor. Furthermore, households are likely to shift their consumption toward consumer durables and other goods and services that are not all that cheap compared to their prices abroad. These factors will lead to the real effective *lowering* of the purchasing power parity exchange rate, which will reduce the rate of growth in dollar GDP below that of GDP in domestic currency.

Because of the deficiencies of both exchange rates for comparing economic output available for military spending and because of these two contrary trends in exchange rates, we chose to measure future Chinese output in dollars using a combination of both exchange rates, explicitly projecting likely future changes in both the market and purchasing power parity exchange rates. Using this combination of what we argue are realistic future market and PPP exchange rates, we project that by 2025 China's GDP will run $9.45 trillion in 2001 dollars. At that time, China's economy will be slightly smaller than that of the U.S. economy today. Assuming the U.S. economy grows at an average annual rate of 3 percent, in 2025 China's economy would be about half the size of the U.S. economy.

Budgetary Pressures

China is becoming older, wealthier, and more urbanized. These changes are generating pressures for additional government spending on pensions, health care, public infrastructure, and the environment. Although China's economy will continue to grow in the coming decades, the ability of the Chinese government to raise the share of output channeled into military spending will be tightly constrained because of these competing demands for government spending on social benefits and other nonmilitary expenditures.

China's current pension system does not cover rural inhabitants or the many urban dwellers who work in the rapidly growing nonstate sector. In addition, many state-owned enterprises and municipalities, the organizations responsible for making pension payments, are facing financial difficulties; in some cases enterprises have gone bankrupt. Expanding China's pension system to cover the entire population with a very modest pension and having the central government cover the pension obligations of those institutions that are unable to meet their commitments will be expensive. We estimate the cost of a modest universal pension system would rise from 3.1 percent of GDP in 2005 to 5.1 percent in 2025.

If China aspires to sustained economic growth and regional leadership, it will require massive catch-up investments in secondary and tertiary education. As Thailand has discovered, rapid economic growth can proceed to a certain point with limited mass education but then hits a ceiling. Understanding this, the Chinese government has committed itself to raising expenditures on education and research and development more rapidly than overall growth in government spending.

China's publicly financed health care system is also in bad shape because government funding has failed to keep pace with either rising demand or the costs of delivering health care services. Especially in rural areas, the Chinese, pay a large share of their total health costs out of their own pockets. The Chinese government is under increasing pressure to make up deficiencies in the health care system. In light of these pressures and based on commitments made by Chinese officials, we project that future government expenditures on education and health as a share of GDP will rise from 4.9 percent of GDP in 2000 to 8.5 percent in 2025.

The Chinese government faces a number of unrecognized liabilities, the most important of which are bad debts owed the state-owned banking sector by state-owned enterprises. We estimate China's total net debt, once these obligations are recognized, is on the order of 60 percent of GDP. When these obligations are recognized, they will result in a one-time increase in China's national debt. As financial markets are liberalized, competition for deposits will push up interest rates, forcing the government to pay market rates on its domestic bonds. Consequently, the Chinese government will face a substantial increase in interest costs in the coming decades.

Combined, spending on pensions, education and health, and increased interest charges are projected to rise from 7.5 percent of GDP in 2000 to 15.5 percent in 2025, an increase of 8 percentage points. These new expenditures are projected to raise total government spending from an estimated 23 percent of GDP in 2003 to a maximum of 33 percent of GDP by 2025. The projected shares of government expenditures in GDP are quite high compared to other middle-income developing countries. The average for this group is 23.1 percent of GDP. Thus, substantially raising government expenditures as a share of GDP will not be an easy task.

Current Military Budgets

China's military budgets are a visible manifestation of national strategic intentions, priorities, and policies. In this respect, trends in the budget as a percentage of GDP are important. The relative scale and dynamism of spending are also a reflection of the state of civil-military relations. Properly used, estimates of Chinese military spending may even be employed as a supplementary metric of overall military capabilities.

However, the official Chinese defense budget includes only a portion of the total defense budget. Although it includes most personnel, operations and maintenance, and equipment costs, it excludes

- foreign weapons procurement
- expenses for paramilitaries (People's Armed Police)
- nuclear weapons and strategic rocket programs
- state subsidies for the defense-industrial complex
- some defense-related research and development
- extra-budget revenue *(yusuanwai)*.

We have reestimated China's defense budget by including estimates of these omitted items. To the official budget of RMB 185.3 billion in 2003 ($22.4 billion), we have added estimates of Chinese imports of military equipment ($3.6 billion), provincial support to national defense ($1.18 billion), and paramilitary expenses ($3 billion). Finally, we assumed that defense-industrial subsidies and R&D funding are bounded by the totals listed in the national budget and could not exceed $3.1 billion and $4.3 billion, respectively. Using this combination of data and assumptions, we estimate total defense expenditures ran between $31 billion and $38 billion in 2003, or 1.4 to 1.7 times the official number.

China's Defense Industries

Over the past 20 years, one of the most prominent and consistent conclusions drawn from research on China's defense industrial complex has been that China's defense production capabilities are rife with weaknesses and limitations. We find, however, that China's defense industries are undergoing substantial change. China has a large, if technologically backward, defense-manufacturing base. Over the past decade, military equipment manufacturers have begun to produce a wide range of increasingly advanced weapons. China also has a growing pool of technical talent in its civilian sector from which Beijing is now attempting to draw for work in the defense sector. The government is also making a concerted effort to reform the institutional framework and incentives under which the defense industry operates. These reforms are taking time, but a number of initial indications suggest that progress is occurring, especially compared with the previous rounds of rather feckless attempts at defense industry reform that have occurred since economic reform was initiated in 1978. In the words of General Li Jinai, the head of the General Armaments Department, the lead defense procurement agency, "there has been a marked improvement in national defense scientific research and in building of weapons and equipment. The past five years has been the best period of development in the country's history."[1]

One of the primary reasons for the slow technological progress in China's defense industries in the 1980s and 1990s was a lack of incentives for innovation. For example, China's defense manufacturers were paid cost plus five percent for the weapons and equipment they produced. Decisions about which company would produce a particular item were made by administrative fiat and ministerial bargaining, rather than through competitive bidding among manufacturers. As a result, military equipment producers had little financial interest in improving the quality of the weapons systems they produced or the efficiency with which they manufactured or designed them, because improvements had little effect on the orders the company received or the profits it made.

Beginning in spring 1998, during the 9th National People's Congress meeting, China's leadership initiated a new series of actions to

---

[1] Wang Wenjie, "Delegate Li Jinai Emphasizes: Grasp Tightly the Important Strategic Opportunity, Accelerate the Development By Leaps of Our Army's Weapons and Equipment," *Jiefangjun Bao,* 8 March 2003, p. 1.

reform the operation of the defense procurement system at the government level and to restructure the defense industries at the enterprise level. Part of the reform was organizational: The government broke the defense industry "companies" into semiautonomous enterprises able to compete with each other. It also introduced some bidding for contracts. In addition, the government initiated a "grand strategy" for improving the technological capabilities of China's defense industries. This strategy has three main elements:

1. *Selective modernization.* China's leaders have spoken of exploiting China's strength in aerospace, the manufacturing of missiles, and electronics technology or concentrating on C4ISR, accurate strike weapons, and other crucial high-tech equipment.
2. *Civil-military integration.* The government has attempted to provide incentives and make organizational changes to capture for military enterprises the improvements in efficiency and technological sophistication that state-owned enterprises in computing, shipbuilding, and electronics have made in production for civilian clients.
3. *Exploiting advanced foreign technology.* Given the backwardness of China's defense industries relative to the advanced nations of the world, the Chinese see the best way to achieve self-sufficiency as involving the importation of the technology needed to enable China's defense industries to produce state-of-the-art military equipment.

These strategies appear to have had some success in the military information technology, shipbuilding, and aerospace industries. On the civilian side, these industries are manufacturing globally competitive products. A number of new weapons systems, including Chinese destroyers, missiles, and C4ISR systems, have shown marked improvements over past production by incorporating these technologies. In contrast, China's military aviation industry continues to systematically underperform most of China's other defense industrial sectors. Its diversification into commercial production of aviation and nonaviation products has not significantly contributed to the modernization of the industry. In this sector, the PLAAF has had to rely on imports of planes or key components to obtain aircraft that are even somewhat competitive with those flown by the United States.

Many of the weaknesses of China's defense industrial sector could be overcome in the short to medium term, assuming China does not deviate from its present course of reform and continues to invest in defense production. If the government continues to push for open contracting and takes a tough line on cost overruns, the rate of innovation and quality of weapons systems should continue to improve. China possesses a large and growing pool of technical talent that could be convinced to work for the defense sector if provided with the proper incentives. However, even though reform could be accelerated, it will not happen overnight. It will take time to train new employees into skilled defense industry engineers and technicians. It will also take time to change management behavior and stimulate innovation, even after new management incentive systems are implemented.

Future Military Budgets

In Chapter Seven we provided two sets of projections of potential Chinese military expenditures through 2025 in 2001 dollars. The high-end forecast was based on our assumption that the maximum share of output that the Chinese government would be able to spend on defense in the context of current relatively benign perceptions of external threats would be 5.0 percent of GDP. The second forecast is a mid-range projection based on the assumption that military spending will not rise above 2.3 percent of GDP. Projections were made by major expenditure category in renminbi and then converted into 2001 dollars using projected market or PPP exchange rates, whichever was more appropriate. For example, personnel costs were converted into 2001 dollars using PPP exchange rates while procurement costs were converted at market exchange rates. We believe that this composite approach provides the most accurate comparison between Chinese and U.S. military expenditures.

Both projections yield substantial sums. By 2025, our mid-range projection yields spending of $185 billion. However, 44 percent of these expenditures consist of personnel costs: operations and maintenance and procurement and RDT&E costs were projected at $52 billion and $51 billion, respectively. The projection of military spending under the maximum expenditure scenario results in considerably higher numbers: military spending rises from an estimated $76 billion in 2003 to $403 billion in 2025, at which time China would be spending close to a third more than the United States did in 2003. However, this projection is truly a maximum in terms of what China is likely to be able to afford. It is based on the assumption that the Chinese leadership would be willing to raise military expenditures to 5 percent of its GDP over a period when political pressures to increase spending on health, education, and pensions—not to mention infrastructure, the environment, and unemployment assistance—will be very strong.

We also projected the potential resources that China may devote to purchasing military assets in the coming two decades. To provide a better sense of the cumulative effect on force structure of the projected defense expenditures on procurement, we compared the projected cumulative totals spent on procurement and research and development through 2025 with U.S. expenditures on these items in 2001 dollars over the past 22 years. Military capabilities, especially stocks of weaponry, are the result of cumulative spending over time, not just current spending. This exercise provides a measure of what China may spend.

Procurement was appreciable in the high-end case. The cumulative total would be close to half of what the United States spent on military procurement and RDT&E between 1981 and 2003. Under this scenario, no other country outside the United States would rival China in terms of weapons stocks. In the mid-range projection, procurement spending is still appreciable, but in this case, even after 22 years, China's cumulative expenditures on procurement would be only 22.0 percent of what the United States spent between 1981 and 2003.

Finally, to give the United States Air Force a more tangible measure of what potential future Chinese expenditures might be on air assets, we provide notional projections of future Chinese spending on research and development and procurement of air assets and compare them to past USAF expenditures in the same categories. The magnitudes are revealing. In our view, the maximum likely expenditures that China would make on RDT&E and procuring weapons and equipment for the PLAAF between 2003 and 2025 would be on the order of

$490 billion. This is a large sum of money. However, the cost in 2001 dollars of the current USAF inventory of weapons and equipment and the associated RDT&E to develop these systems is more than twice this number. Using our mid-range projection of military expenditures, PLAAF expenditures on procurement and RDT&E would run $229 billion, 22.0 percent of cumulative USAF expenditures over the past 22 years on the same categories.

In short, China will have the economic wherewithal to substantially increase military expenditures in the coming two decades. When converted using an appropriate combination of purchasing power parity and market exchange rates to reflect differences in purchasing power between China and the United States, Chinese military expenditures measured in 2001 dollars are likely to run on the order of $185 billion in 2025. In our maximum expenditure scenario, they could run as high as $403 billion in 2001 dollars in 2025, a third more than the 2003 U.S. defense budget.

How Can We Tell If China Is Straying from the Projected Course?

These projections of future economic output and military expenditures are designed to assist in sizing future threat levels. Because of the many unknowns involved, no forecasts can be precisely accurate. Rather, our projections provide a logical, internally consistent framework in which to assess likely future directions in China's military spending.

Throughout our research for this study, we have identified and utilized a variety of indicators to assess current and project likely future expenditure levels. Many of these indicators are already being watched by Air Force intelligence analysts. Below, we have culled a subset of these indicators that we believe are most salient for tracking likely future changes in China's overall military capabilities. They will help detect whether China is moving in the direction forecast or whether future military expenditures and capabilities are likely to diverge substantially, up or down, from those projected here.

Indicators of Economic Growth

Growth Rates of GDP and Key Economic Sectors. The future size of China's economy will be a major determinant of military spending. We have argued that the rate of growth in economic output is destined to slow, most likely to an average annual rate of 5 percent per year through 2025. But China may surprise us. After slowing to a range of 7 to 8 percent per year between 1998 and 2002, official statistics show growth of 9.1 percent again for 2003. Even if one assumes that the official numbers are inflated by 1 to 1.5 percentage points, sustained growth at the 2003 pace for more than a decade would result in a much larger economy and hence a much larger resource base for military spending than in our projections. Our projected slowdown in growth is predicated on a slowdown in increases in output of key economic sectors, especially industry and exports. If growth in these sectors is substantially more or less than projected, overall economic growth will also diverge from our forecasts. Consequently, we advocate utilizing the following indicators to determine whether growth is following the projected path:

- Three consecutive years of growth in GDP of more than 2.5 percentage points or less than 2.5 percentage points compared with our projected rates
- Three consecutive years of growth in industrial output of more than 3 percentage points or less than 3 percentage points compared with our projected rates
- Three consecutive years of growth in exports of over 10 percent or under 5 percent.

Foreign Direct Investment (FDI). Inflows of foreign direct investment into China have been very strong for the past decade, averaging well over $35 billion a year. Our base case forecast assumes continued inflows, but at stable levels, not continued growth. If FDI were to fall sharply or to run at much higher levels, the outlook for growth would be quite different, depending on the direction of change. To track the direction of FDI, we use the following indicator:

- Net inflows of FDI based on balance of payments data of over $50 billion or less than $25 billion for three consecutive years.

The Financial System. The financial system is a looming threat to economic growth in China. If nonperforming loans are not written off in a timely manner and if credit allocation is not improved, China is likely to experience an overt banking crisis that could result in a sharp slowdown in growth. Unfortunately, China's financial data, especially concerning credit quality, are very deficient. Consequently, the health of China's financial system needs to be tracked both by financial sector indicators and by watching government policies and institutional changes within the financial sector. For example, if the government makes serious moves to restructure or sell off the state-owned banks, credit allocation is likely to improve, along with China's economic outlook. The following indicators are of use in assessing this sector:

- Very substantial year-on-year changes in official data on nonperforming assets
- The total cost of government programs to bail out banks that are not creditworthy
- Moves to privatize, split up, or restructure the four largest state-owned banks.

Changes in Agricultural Policy. The current projection assumes that China will slowly improve agricultural policies to the point where deficiencies in agriculture do not threaten growth. However, if the government fails to improve land tenure arrangements, moves away from targeting grain output, or fails to reduce tax burdens on rural peasants, agricultural performance would be worse than expected with a corresponding effect on growth. Therefore, analysts of the Chinese economy need to pay particular attention to agricultural policy:

- Announcements and reports on implementation of land titling and transfer; the creation of markets and free operation of markets for agricultural land; inputs, including water; and output, including key crops

- Changes in the amount of cultivable land
- Availability of water for irrigation
- The harvest and yields for major crops (rice, wheat, soybeans)
- Changes in the rate of growth in output of crops and animal products.

Budget Indicators

We have argued that military spending will be constrained by increasing demands for government social spending on pensions, education, and health; increased interest costs on government debt; and the need to invest in infrastructure and to rectify environmental damage. We also argue that the Chinese government's ability to increase the share of output taken by taxes and fees is limited in light of the extent of corruption and fraud in the tax system. Our mid-range projection that Chinese military spending will not exceed 2.3 percent of GDP is predicated on our view that demands for social spending and Chinese perceptions of relatively limited threats to national security will make it difficult to raise military spending much above this level. In contrast, our maximum projection assumes that threat perceptions will be such that China will be able to raise spending to 5 percent of GDP.

The following programmatic and statistical indicators should provide early indications whether the Chinese government will be able to raise projected revenues and whether it is choosing to respond to pressures for greater social spending:

- Overall growth in total government revenues from taxes and fees of more than 4 percentage points faster or 2 percentage points slower than growth in GDP for three consecutive years
- Growth in overall government expenditures (central and provincial) of more than 4 percentage points faster or 2 percentage points slower than GDP for three consecutive years
- Increases in the shares of government spending on pensions, education, and health of more than 1 percentage point for three consecutive years
- Introduction of a universal national pension program
- Introduction and implementation of national health reform

- Implementation of plans for large-scale investments in diverting water from the south to the north
- Implementation of an accelerated national highway construction program.

Indicators of Improvements in the Defense Industries

China has the basics needed to manufacture sophisticated weapons systems. It has large numbers of well-trained engineers and technicians and an increasingly sophisticated manufacturing base for civilian goods. Recent procurement budgets in the overall defense budgets provide substantial sums of money for the purchase of military equipment and weapons. To date, however, China has not had great success in putting these pieces together. Designs, especially in aviation, remain outmoded. Key components remain substandard, even in relatively successful areas such as shipbuilding.

Traditional intelligence analysis such as observed measurements of performance, analysis of drawings, and engineering assessments of Chinese-produced items will remain the most important means of evaluating the ability of China's defense industries to produce more sophisticated systems. However, these measurements come after the fact. A number of leading indicators should help intelligence analysts evaluate whether China's defense industries are becoming more capable of producing sophisticated, quality weapons systems:

- Reports that traditional producers are losing major contracts through a competitive bidding process and evidence that production has been transferred to the winning bidder
- Credible reports of substantial rewards or penalties for producing superior or inferior products
- Closure of poorly performing plants whereas better performing plants continue to operate
- Significant contract awards to nontraditional suppliers, including nonstate enterprises
- Divestures and acquisitions driven by decisions taken by enterprise management, not ministries

- Privatization of defense manufacturers
- Substitution of domestic production for imports.

Indicators of Changes in Military Spending

We have argued above that the official budgets and spending from some clearly identifiable nondefense budget categories capture most of China's spending on the military. Although spending may be shifted from one section of the budget to another, over the past few decades trends in the official budget and other categories that we have identified have been strongly correlated with other indicators of resources going to the military. Thus, these spending categories have provided useful indications of the size and trends in resources the Chinese government has provided the military. These budgets can also be used to highlight future trends in expenditures. To ascertain whether our mid-range projection remains plausible, the following indicators should be of use:

- Changes in official budgetary expenditures in nominal terms
- Changes in official budgetary expenditures in real terms deflated by the GDP deflator
- Changes in the share of GDP accounted for by official military spending
- Changes in expenditures on research and development
- Reports of imports of weapon systems by type, volume, and value
- Changes in the allocation of military expenditures as reported in the biannual *Defense White Papers.*

Bibliography

2000–2001 *nian zhanlue pinggu* [2000–2001 Strategic Assessment], Beijing, China: Junshi kexue chubanshe, 2000.

Agarwala, Ramgopal, *China: Reforming Intergovernmental Fiscal Relations,* Washington D.C.: World Bank Discussion Paper 178, 1991.

Allen, Kenneth W., Glenn Krumel, Jonathan D. Pollack, *China's Air Force Enters the 21st Century,* Santa Monica, Calif.: RAND Corporation, MR-580-AF, 1995.

Arnett, Eric, "Military Technology: The Case of China," *SIPRI Yearbook 1995: Armaments, Disarmament and International Security,* New York: Oxford University Press, 1995.

Baark, Erik, "Fragmented Innovation: China's Science and Technology Policy Reforms in Retrospect," in Joint Economic Committee, ed., *China's Economic Dilemmas in the 1990s: The Problems of Reforms, Modernization, and Interdependence,* Washington, D.C.: U.S. Government Printing Office, 1991.

———, "Military Technology and Absorptive Capacity in China and India: Implications for Modernization," in Eric Arnett, ed., *Military Capacity and the Risk of War: China, India, Pakistan and Iran,* Oxford, England: Oxford University Press, 1997.

Bitzinger, Richard A., "Arms To Go: Chinese Arms Sales to the Third World," *International Security,* Fall 1992.

———, "Military Spending and Foreign Military Acquisitions by the PRC and Taiwan," in James R. Lilley and Chuck Downs, eds., *Crisis in the Taiwan Strait,* Washington D.C.: National Defense University Press, 1997, p. 77, Figure 2.

_____, "Just the Facts, Ma'am: The Challenge of Analysing and Assessing Chinese Military Expenditures," *China Quarterly*, No. 173, 2003.

Bitzinger, Richard A., and Chong-Pin Lin, *The Defense Budget of the People's Republic of China*, Washington, D.C.: Defense Budget Project, 1994.

Brömmelhörster, Jorn, and John Frankenstein (eds.), *Mixed Motives, Uncertain Outcomes: Defense Conversion in China*, Boulder, Colo.: Lynne Rienner Publishers, 1997.

Cao Haili, "The Chinese Army Has Sailed Out of the Business Sea," *Caijing* [Finance and Economics], January 1999.

China Academy of Military Sciences Editing Group (ed.), *Zhongguo junshi caiwu shiyong daquan* [Practical Encyclopedia of Chinese Military Finance] (ZJCSD), Beijing: Jiefangjun chubanshe, 1993.

China Finance Yearbook 2002, Beijing: Ministry of Finance, 2002.

China Statistical Yearbook (CSY), Beijing: China Statistics Press, 1978–2002.

China's National Defense in 1998, Beijing: State Council Information Office, December 1998.

China's National Defense in 2000, Beijing: State Council Information Office, December 2000.

China's National Defense in 2002, Beijing: State Council Information Office, December 2002.

Chinese Military Encyclopedia Editing Group (ed.), *Jundui houqin fence* [Military Logistics Volume], Beijing: Junshi kexueyuan chubanshe, 1985.

Chow, Gregory, and An-loh Lin, "Accounting for Economic Growth in Taiwan and Mainland China: A Comparative Analysis," *Journal of Comparative Economics*, Vol. 30, No. 3, September 2002.

Ci Shihai, *Budui Zhuangbei Guanli Gailun* [Army Equipment Management Theory], Beijing: Junshi kexue chubanshe, 2001.

Conroy, Richard, *Technological Change in China*, Paris: Development Centre of the Organization for Economic Co-operation and Development, 1992.

Contemporary China Series Editing Group (ed.), *Dangdai Zhongguo jundui de houqin gongzuo* [Contemporary Chinese Military Logistics Work], Beijing: Zhongguo shehui kexue chubanshe, 1990.

Council on Foreign Relations, Independent Task Force Report, *Chinese Military Power*, Washington, D.C., 2003.

Ding, Arthur S., "China Defense Finance: Content, Process and Administration," *The China Quarterly*, No. 146, June 1996a.

———, "Economic Reform and Defence Industries in China," in Gerald Segal and Richard S. Yang, ed., *Chinese Economic Reform*, New York: Routledge, 1996b.

Eikenberry, Karl W., *Explaining and Influencing Chinese Arms Transfers*, McNair Papers 36, Washington, D.C.: National Defense University, February 1995.

Energy Information Agency, *International Energy Outlook 2002*, U.S. Department of Energy, Washington, D.C., March 2002.

Forney, Matt, "A Chinese Puzzle: Unwinding Army Enterprises," *Wall Street Journal*, December 15, 1998.

Frankenstein, John, "China's Defense Industries: A New Course?" in James C. Mulvenon and Richard H. Yang, eds., *The People's Liberation Army in the Information Age*, Santa Monica, Calif.: RAND Corporation, 1999.

———, "The People's Republic of China: Arms Production, Industrial Strategy and Problems of History," in Herbert Wulf, ed., *Arms Industry Limited*, Solna, Sweden: Stockholm Institute for Peace Research International, 1993.

Frankenstein, John, and Bates Gill, "Current and Future Challenges Facing Chinese Defense Industries," *The China Quarterly*, June 1996.

Frieman, Wendy, "Arms Procurement in China: Poorly Understood Processes and Unclear Results," in Eric Arnett, ed., *Military Capacity and the Risk of War: China, India, Pakistan and Iran*, Oxford, England: Oxford University Press, 1997.

———, "China's Military R&D System: Reform and Reorientation," in eds., Denis Fred Simon and Merle Goldman, eds., *Science and Technology in Post-Mao China*, Cambridge, Mass.: Harvard University Press, 1989.

———, "China's Defence Industries," *The Pacific Review*, Vol. 6, No. 1, 1993.

Fung, K.C., Hitoma Iizaka, and Stephen Parker, "Determinants of U.S. and Japanese Direct Investment China, *Journal of Comparative Economics*, Vol. 30, No. 3, September 2002.

Gallagher, Joseph P., "China's Military Industrial Complex," *Asian Survey,* Vol. XXVII, No. 9, September 1987.

Gao Dianzhi, *Zhongguo guofang jingji guanli yanjiu* [Research on Chinese Defense Economic Management], Beijing: Junshi kexueyuan chubanshe, 1991.

Gill, Bates R., *Chinese Arms Transfers: Purposes, Patterns and Prospects in the New World Order,* Westport, Conn.: Praeger Publishers, 1992.

_____, "Chinese Defense Procurement Spending: Determining Intentions and Capabilities," in James R. Lilley and David Shambaugh, eds., *China's Military Faces the Future,* Washington, D.C.: American Enterprise Institute, 1999.

_____, "The Impact of Economic Reform on Chinese Defense Production," in C. Dennison Lane, ed., *Chinese Military Modernization,* London: Paul Kegan International, 1996.

_____, "Chinese Military Technical Developments: The Record From Western Assessments, 1979–1999," as published in James C. Mulvenon and Andrew N.D. Yang, *Seeing Truth from Facts,* Santa Monica, Calif.: RAND Corporation, 2001.

Gill, Bates, and Lonnie Henley, *China and the Revolution in Military Affairs,* Carlisle, Pa.: Strategic Studies Institute, 1996.

Grimmett, Richard, *Conventional Arms Transfers to Developing Nations, 1994–2001,* Washington, D.C.: Congressional Research Service, 2002.

Harrold, Peter, and Rajiv Lall, *China: Reform and Development in 1992–1993,* World Bank Discussion Paper #215, Washington, D.C.: The World Bank, 1993.

Holz, Carsten A., "The Changing Role of Money in China and Its Implications," *Comparative Economic Studies,* Vol. XLIII, No. 3, Fall 2000.

Hu Angang, "Creative Destruction and Restructuring: China's Urban Unemployment and Social Security, (1993–2000)," mimeo, Harvard Kennedy School of Government, 2002.

Institute of International Security Studies (IISS), "China's Military Expenditures," *The Military Balance 1995/96,* London: IISS, 1995.

_____, *The Military Balance, 1996–1997,* London: IISS, 1997.

_____, *The Military Balance, 2002–2003,* London: IISS, 2003.

International Monetary Fund (IMF), "China's Medium Term Fiscal Challenges," *World Economic Outlook*, Washington, D.C., April 2002.

_____, *International Financial Statistics*, Washington, D.C., various years.

Janes Armour and Artillery, 2002 (accessed via Janes Online, 9 August 2002).

Jencks, Harlan, "The General Armaments Department," in James C. Mulvenon and Andrew N.D. Yang, *The People's Liberation Army as Organization: Reference Volume v1.0*, Santa Monica, Calif.: RAND Corporation, 2002.

Jin Songde et al., *guofang jingji lun* [On National Defense Economics], Beijing: Jiefangjun chubanshe, 1987.

Kravis, Irving B., "An Approximation of the Relative Real Per Capita GDP of the People's Republic of China, *Journal of Comparative Economics*, 1981, Vol. 5, No. 1, pp. 60–78.

Ku Guisheng and Quan Linyuan, *Junfeilun* [On Military Budgets], Beijing: National Defense University, 1999.

Lane, C. Dennison (ed.), *Chinese Military Modernization*, London: Paul Kegan International, 1996.

Lardy, Nicholas R., *Integrating China into the Global Economy*, Washington, D.C.: Brookings Institution Press, 2002.

Lawrence, Susan, and Bruce Gilley, "Bitter Harvest," *Far Eastern Economic Review*, April 29, 1999.

Li Jiamo, "The Design of Primary Aircraft Components Should Be Closely Integrated with Manufacturing Enterprises," *Guofang Keiji Gongye*, National Defense Science and Technology Industry, 2002, No. 5.

Lieberthal, Kenneth, and Michel Oksenberg, *Policy Making in China: Leaders, Structures, and Processes*, Princeton, N.J.: Princeton University Press, 1988.

Lilley, James R., and David Shambaugh (eds.), *China's Military Faces the Future*, Washington, D.C.: American Enterprise Institute, 1999.

Lin, Chong-Pin, *The Defense Budget of the People's Republic of China*, Washington, D.C.: Defense Budget Project, 1994.

Lin Shuanglin, "China's Government Debt: How Serious?" *China: An International Journal*, I, March 2003.

Lin Yichang and Wu Xizhi, *Guofang jingjixue jichu* [Basic Defense Economics], Beijing: Junshi kexueyuan chubanshe, 1991.

Lin Yi-min and Tian Zhi, "Ownership Restructuring in Chinese State Industry: An Analysis of Evidence on Initial Organizational Changes," *China Quarterly,* September 2001.

Liu Zhiqiang, "Foreign Direct Investment and Technology Spillover, *Journal of Comparative Economics,* Vol. 30, No. 3, September 2002.

Long Youcai, and Wang Zong (eds.), *Jundui caiwu jianshe* [Military Economic Construction], Beijing: Jiefangjun chubanshe, 1996.

Lu Zhuhao (ed.), *Zhongguo junshi jingfei guanli* [China's Military Budget Management], Beijing: Jiefangjun chubanshe, 1995.

Ma Guonan and Ben S.C. Fung, "China's Asset Management Corporations," BIS Working Paper N. 115, Basle: Bank for International Settlements, August 2002.

Medeiros, Evan S., "Revisiting Chinese Defense Conversion: Some Evidence from China's Shipbuilding Industry," *Issues and Studies,* May 1998.

Medeiros, Evan S., and Bates Gill, *Chinese Arms Exports: Policy, Players, and Process,* Strategic Studies Institute, Carlisle, Pa.: U.S. Army War College, 2000.

Mulvenon, James C., *Chinese Military Commerce and U.S. National Security,* Santa Monica, Calif.: RAND Corporation, 1997a.

_____, *Professionalization of the Senior Chinese Officer Corps: Trends and Implications,* Santa Monica, Calif.: RAND Corporation, MR-901-OSD, 1997b.

Mulvenon, James C., and Andrew N.D. Yang (eds.), *Seeking Truth from Facts: A Retrospective on Chinese Military Studies in the Post-Mao Era,* Santa Monica, Calif.: RAND Corporation, CF-160-CAPP, 2001.

Mulvenon, James C., and Richard H. Yang (eds.), *The People's Liberation Army in the Information Age,* Santa Monica, Calif.: RAND Corporation, CF-145-CAPP/AF, 1999.

National Defense University Development Institute (ed.), *Zhongguo guofang jingji fazhan zhanlue yanjiu* [Research on China's National Defense Economic Development Strategy], Beijing: Guofang daxue chubanshe, 1990.

Naughton, Barry, *Growing Out of the Plan,* Cambridge, United Kingdom: Cambridge University Press, 1996.

Nyberg, Albert, and Scott Rozelle, *Accelerating China's Rural Transformation,* Washington, D.C.: The World Bank, August 1999.

Ostrov, Benjamin A., *Conquering Resources: The Growth and Decline of the PLA's Science and Technology Commission for National Defense,* Armonk, N.Y.: M.E. Sharpe, 1991.

Overholt, William H., *The Rise of China: How Economic Reform Is Creating a New Superpower,* New York: W.W. Norton and Company, 1993.

People's Liberation Army Logistics College Technology Research Section, ed., *Junshi houqin cidian* [Military Logistics Dictionary], Beijing: Jiefangjun chubanshe, 1991.

Pillsbury, Michael, *China Debates the Future Security Environment,* Washington, D.C.: National Defense University Press, 2000.

_____, *Chinese Views of Future Warfare,* Washington, D.C.: National Defense University Press, 1998.

Prime, Penelope B., "Taxation Reform in China's Public Finance," Joint Economic Committee of Congress, U.S. Printing Office, April 1991.

Ren Ruone and Chen Kai, "China's GDP in U.S. Dollars Based on Purchasing Power Parity," Policy Research Working Paper #1414, Washington, D.C.: The World Bank, January 1995.

Ryten, Jacob, "Evaluation of the International Comparison Project," United Nations Economic and Social Council, Statistical Commission, E/CN.3/1999/8, November 16, 1998.

Shambaugh, David, *Modernizing China's Military: Progress, Problems, and Prospects,* Berkeley: University of California at Berkeley Press, 2003.

_____, "Wealth in Search of Power: The Chinese Military Budget and Revenue Base," paper delivered to the Conference on Chinese Economic Reform and Defense Policy, Hong Kong, July 1994.

Sidhu, Waheguru Pal Singh, and Jing-dong Yuan, *China and India: Cooperation or Conflict,* Boulder, Colo.: Lynne Rienner Press, 2003.

Singh, Ravinder Pal (ed.), *Arms Procurement Decision Making: China, India, Israel, Japan, South Korea and Thailand,* Stockholm International Peace Research Institute, Oxford, United Kingdom: Oxford University Press, 1998.

Sun Zhenyuan, *Zhongguo guofang jingji jianshi* [China's National Defense Economic Construction], Beijing: Junshi kexueyuan chubanshe, 1991.

Suttmeier, Richard P., "China's High Technology: Programs, Problems, and Prospects," in Joint Economic Committee, ed., *China's Economic Dilemmas in the 1990s: The Problems of Reforms, Modernization, and Interdependence,* Washington, D.C.: U.S. Government Printing Office, 1991.

_____, "Emerging Innovation Networks and Changing Strategies for Industrial Technology in China: Some Observations," *Technology in Society,* Vol. 19, No. 3 & 4, 1997.

U.S. Department of Defense, *Annual Report on the Military Power of the People's Republic of China,* Report to Congress Pursuant to the FY2000 National Defense Authorization Act, Washington, D.C.: Department of Defense, July 28, 2002.

U.S. Department of Defense, *Annual Report on the Military Power of the People's Republic of China,* Report to Congress Pursuant to the FY2000 National Defense Authorization Act, Washington, D.C.: Department of Defense, July 28, 2003.

U.S. Department of Defense, *Annual Report on the Military Power of the People's Republic of China,* Report to Congress Pursuant to the FY2000 National Defense Authorization Act, Washington, D.C.: Department of Defense, May 2004.

U.S. Department of State, Bureau of Verification and Compliance, *World Military Expenditures and Arms Transfers 1998,* Washington, D.C.: U.S. Department of State, 2000.

U.S. General Accounting Office, *Export Controls: Sale of Telecommunications Equipment to China* (GAO/NSIAD-97–5), November 1996a.

_____, *Export Controls: Sensitive Machine Tool Exports to China* (GAO/NSIAD-97–4, November 1996b).

Wang Hongqing and Zhang Xingye (eds.), *Zhanyixue* [The Science of Campaigns], Beijing: Guofang Daxue, May 2000.

Wang Qincheng and Li Zuguo (eds.), *Caiwu daquan* [Finance Encyclopedia], Urumqi: Xinjiang renmin chubanshe, 1993.

Wang Shaoguang, "The Military Expenditure of China, 1989–98," *SIPRI Yearbook 2000,* Oxford: Oxford University Press, 2000.

Wang Wenrong (ed.), *Zhanluexue* [The Science of Strategy], Beijing: Guofang daxue chubanshe, 1999.

Wolf, Charles, Jr., Anil Bamezai, K.C. Yeh, and Benjamin Zycher, *Asian Economic Trends and Their Security Implications,* Santa Monica, Calif.: MR-1143-OSD/A, 2000.

Wolf, Charles, Jr., K.C. Yeh, Benjamin Zycher, Nicholas Eberstadt, Sung-Ho Lee, *Fault Lines in China's Economic Terrain,* Santa Monica, Calif.: RAND Corporation, MR-1686, 2003.

Wong, Christine, "China: Provincial Expenditure Review," World Bank briefing, Workshop on Decentralization and Intergovernmental, Washington, D.C., May 13–15, 2002.

_____, "China: Provincial Expenditure Review: Executive Summary," World Bank, Draft of December 11, 2001.

World Bank, *2003 Development Indicators, Economic Outlook,* Washington, D.C., January 2004, http://www.worldbank.org/data/.

_____, *China 2020: Development Challenges in the New Century,* Washington, D.C.: The World Bank, 1997.

_____, *China: National Development and sub-National Finance: A Review of Provincial Expenditures,* Report No. 22951-CHA, Washington, D.C.: The World Bank, April 9, 2002.

_____, *China: Weathering the Storm and Learning the Lessons,* Country Economic Memorandum, Report No. 18768 CHA, Washington, D.C.: The World Bank, May 25, 1999.

_____, *Country Report: China,* Washington, D.C.: The World Bank, 1996.

_____, *World Development Indicators 2002,* CD-ROM version.

Xiang Huaicheng, Minister of Finance, "Report on the Implementation of the Central and Local Budgets for 2002 and the Draft Central and Local Budgets for 2003," 10th National People's Congress: Chinese People's Political Consultative Congress, March 6, 2003. http://www.china.org.cn/english/features/59352.htm.

Xie Dajun, "The Procurement and Supervision of the Manufacture of Foreign Armaments," *Xiandai Junshi,* August 1999.

Yan Wang and Yao Yudong, *Sources of China's Economic Growth, 1952–99: Incorporating Human Capital Accumulation,* Washington, D.C.: The World Bank, July 2001.

Yan Wang, Dianqing Xu, Zhi Wang, and Fan Zhai, "Transition Cost, Options and Impact of China's Pension Reform," Policy Research Working Paper No. 2555, Washington, D.C.: World Bank, February 2001.

Yuan Jing-dong, "India's Rise after Pokhran-II: Chinese Analyses and Assessments," *Asian Survey,* No. 41, November/December 2001.

Zhang Xulong (ed.), *Junshi jingjixue* [Military Economic Science], Shenyang: Liaoning renmin chubanshe, 1988.

Zhang Zhenlong (ed.), *Junshi jingjixue* [The Science of Military Economics], Shenyang: Liaoning renmin chubanshe, 1988.

Zhou Yuan, "Reform and Restructuring of China's Science and Technology System," in Denis Fred Simon, ed., *The Emerging Technological Trajectory of the Pacific Rim,* Armonk, N.Y.: M.E. Sharpe, 1995.

Zhu Qinglin, *Zhongguo Caijun yu guofang zirenminbi peizhi yanjiu,* Beijing: National Defense University Press, 1999.